Local cattle breeds in Europe

Local cattle breeds in Europe

Development of policies and strategies for self-sustaining breeds

edited by:

Sipke Joost Hiemstra
Centre for Genetic Resources, the Netherlands (CGN), Wageningen University and Research Centre, Lelystad, the Netherlands

Yvette de Haas
Centre for Genetic Resources, the Netherlands (CGN), Wageningen University and Research Centre, Lelystad, the Netherlands

Asko Mäki-Tanila
MTT Agrifood Research, Joikonen, Finland

Gustavo Gandini
Department VSA, Faculty of Veterinary Medicine, University of Milan, Milan, Italy

Wageningen Academic
P u b l i s h e r s

ISBN: 978-90-8686-144-6
e-ISBN: 978-90-8686-697-7
DOI: 10.3921/978-90-8686-697-7

Photos cover:
CGN and INIA

Photos breedcases:
Veeteelt
Kerry Cattle Society
MTT
EURECA Consortium Partners

First published, 2010

© Wageningen Academic Publishers
The Netherlands, 2010

Table of contents

Chapter 1

Chapter 2

Chapter 3

Chapter 4

Chapter 5

Chapter 6

Chapter 7

Chapter 8

List of breed cases

Acknowledgements

This book is the result of Action EURECA 012 AGRI GEN RES 870/2004 (EURECA). The EURECA project 'Towards self-sustainability of EUropean, REgional, CAttle breeds' was undertaken by a Consortium of 10 European partners (10 countries) between 2007 and 2010. The Consortium investigated different factors affecting the sustainability of local breed farming. Through the exchange of experiences and research outcomes across countries, and through interaction with a variety of stakeholders, the Consortium aimed to contribute to the conservation of local cattle breeds in Europe.

The EURECA project received financial support from the European Commission, Directorate-General for Agriculture and Rural Development, under Council Regulation (EC) No 870/2004. The financial support from the EC served as co-funding for the national funding sources of the project. Therefore, the Consortium partners are particularly grateful to the national funding agencies for their important role in supporting the strengthening of conservation policies and strategies for local cattle breeds across Europe, and – last but not least – to the farmers, stakeholders and experts who contributed voluntarily to the project with data and opinions on the topic of the EURECA project.

Contributors

A great many people contributed to this publication. Although this book was written by a limited few, the assistance of many individuals employed by EURECA Consortium partners and other collaborating experts (listed below) was crucial for the development of the project, for data collection and for analysis of the data.

Contributions from EURECA Consortium and EURECA Experts

Centre for Genetic Resources, the Netherlands (CGN)
Wageningen University and Research Centre, Lelystad, the Netherlands
(EURECA coordinator)
- Sipke Joost Hiemstra
- Yvette de Haas
- Rita Hoving
- Lucia Kaal
- Myrthe Maurice-van Eijndhoven
- Debbie Bohte-Wilhelmus
- Jack Windig
- Henri Woelders

Institut de l'Elevage, Paris, France
(EURECA Consortium partner)
- Delphine Duclos
- Laurent Avon
- Xavier Dornier
- Marina Hohl
- Lenaig Menuet
- André Pflimlin

Department VSA, University of Milan, Milan, Italy
(EURECA Consortium partner)
- Gustavo Gandini
- Federica Turri
- Michele Musella

IBBA CNR, Milan, Italy
(EURECA Expert)
- Flavia Pizzi

Université de Liège, Gembloux Agro-Bio Tech, Unité de Zootechnie
Groupe de Génétique et Amélioration animales, Gembloux, Belgium
(EURECA Consortium partner)
- Nicolas Gengler
- Frédéric Colinet
- Elodie Bay

Association Wallonne de l'Elevage, Recherche et Développement, Ciney, Belgium
(EURECA Expert)
- Patrick Mayeres

MTT Agrifood Research, Finland
(EURECA Consortium partner)
- Asko Mäki-Tanilla
- Katriina Soini
- Taina Lilja

Instituto Nacional de Investigacion y Tecnologia Agraria y Alimentaria (INIA), Dpto. Mejora
Genética Animal, Madrid, Spain
(EURECA Consortium partner)
- Clara Diaz
- Jesús Fernández
- Daniel Martín-Collado
- Miguel Toro

Institute of Animal and Aquacultural Sciences, Ås, Norway
(EURECA Consortium partner)
- Theo Meuwissen

National Research Institute of Animal Production, Krakow, Poland
(EURECA Consortium partner)
- Zenon Choroszy

Irish Cattle Breeding Federation (ICBF), Cork Bandon, Ireland
(EURECA Consortium partner)
- Francis Kearney
- Brian Wickham

Estonian Agricultural University, Tartu, Estonia
(EURECA Consortium partner)
- Haldja Viinalass
- Merko Vaga

Charles Darwin University, Darwin, Australia
(EURECA Expert)
- Kerstin Zander

Bioversity International, Rome, Italy
(EURECA Expert)
- Adam Drucker

Contributions from farmers, stakeholders and National Coordinators

In addition to Consortium partners, a variety of stakeholders contributed to the project by providing data and opinions, including FAO National Coordinators for Animal Genetic Resources in Europe, national breed societies/interest groups and individual farmers. Without these valuable contributions the project would not have been possible.

The following National Coordinators (or their alternates) kindly contributed to the Europe-wide survey on local cattle breeds (more details in Chapter 3):
- Beate Berger (Austria)
- Giovanni Bittante (Italy)
- Didier Bouchel (France)
- Andreas Georgoudis (Greece)
- Ladislav Hetényi (Slovakia)
- Sipke Joost Hiemstra (the Netherlands)
- Ante Ivankovic (Croatia)
- Mark Maguire (Ireland)
- Asko Mäki Tanila (Finland)
- Catherine Marguerat (Switzerland)
- Božidarka Marković (Montenegro)
- Elzbieta Martyniuk (Poland)
- Serge Massart (Belgium)
- Vera Matlova (Czech Republic)
- Helle Anette Palmø (Denmark)
- Christos Papachristoforou (Cyprus)
- Isabel García Sanchez (Spain)
- Nina Sæther (Norway)
- Stefan Schulz (Germany)
- Eva-Marie Stålhammar (Sweden)
- Srdjan Stojanovic (Serbia)
- Tamás Szobolevszki (Hungary)
- Luis Telo da Gama (Portugal)
- Haldja Viinalass (Estonia)

Chapter 1

Introduction

Sipke Joost Hiemstra

How can we positively influence the future of our local cattle breeds in Europe? This is the question that experts and researchers from 10 European countries asked themselves in a 3 year project (EURECA), co-funded by the European Commission. Which factors can be identified to contribute to success and which decision-making tools can we suggest for development of breed strategies? With this publication we share the outcome of our studies and suggest recommendations to improve the future of local cattle breeds in Europe for breed managers, policy makers and other stakeholders.

1.1 Why conserve local breeds?

Since the domestication process in the Neolithic Age, livestock has spread all over the world as a result of human migration or interchanges among neighbouring human populations. As they reached different places they slowly adapted to the specific environmental conditions of the area and to the 'cultural' preferences of their new herdsmen, giving rise to the livestock's genetic diversity. In the old days, domestic animals were multifunctional; they were used for draught work, clothes, manure, fuel and food. It was not until 18th century in Europe when these differences between animals within the same species acquired a name, and were called 'breeds'.

After the industrial revolution, the traditional use of domestic animals for draught work, clothes and manure was slowly but steadily substituted by industrial products. With the increasing demand for protein of animal origin, breeds were intensively selected for food purposes and the development of specialised dairy and beef breeds began. This process started at different periods depending upon the country and region. Intensively selected breeds and their high-input high-output production systems have been very successful and widely disseminated, displacing many native breeds which had not undergone any selection process. Luckily, many of the native breeds have survived in areas where high-input high-output systems were not established for economic, cultural or environmental reasons. Native or local breeds are nowadays usually characterised by their limited geographical distribution. Sometimes the expansion is greater, crossing neighbouring regions within a country or even bordering countries, and the breed is then called a regional breed. Throughout this publication we consistently use the term 'local breed'.

Globally, about 20% of all breeds or livestock populations are considered to be 'at risk' and 9% are already extinct (FAO, 2007). Similar figures can be shown for cattle breeds in Europe where at least 130 previously known cattle breeds are already 'extinct' (www.fao.org/dad-is). In terms of numbers of breeds, the majority of the cattle breeds in Europe can be categorised as local breeds. As a large number of breeds became endangered, the 'hidden', previously ignored, values started to be recognised. For some decades now, European society has recognised the important environmental, social, cultural, market and public values of these cattle breeds.

It is time to understand the state of European local cattle populations in order to develop well-oriented policies and strategies for preserving all the values related to the maintenance of cattle genetic diversity in Europe.

Farmer's quote:

'Those who do not understand the old will not understand the new'

1.2 Towards better strategies for the management of local cattle breeds

The most secure conservation strategy is to promote measures that make breeds 'self-sustaining', i.e. breeds that can be maintainted without the need for external economic support, including EU subsidies.

Conservation aims obviously encompass more than just economic independence from subsidies. We should aim for breeds and farming systems capable of maintaining the vigour and the potential to fulfil all conservation aims, including maintenance of genetic variability and, if applicable, the specific cultural, social, economic and environmental values.

From a genetic point of view, the importance of the conservation of between and within breed genetic diversity is widely recognised. Therefore, there is a need to fully integrate proper management of genetic variation in breeding or development programmes for local breeds. Successful breed strategies or policies have to take into account different factors which could have a positive or negative impact on breed survival. It is clear that European countries face similar issues or problems, associated with the conservation and sustainable use of local or regional breeds and with the role these breeds play in rural development and the socio-economic development of agricultural communities. Different - complementary (*in situ* and *ex situ*) and integrated - strategies are needed to conserve local cattle breeds and develop and promote their use.

In the development of breed strategies, a combination of production, market and non-market values should guarantee sufficient profitability of the breed. Many local breeds are already kept for 'multi-functional' reasons; many other breeds could also benefit more from such a strategy. Furthermore, several closely related breeds exist in neighbouring countries or other countries in Europe and enhanced co-operation across countries could help conserve such breeds or breed groups. Therefore, conservation and management of local cattle breeds have trans-national or cross-border dimensions. By sharing experience and knowledge, further co-operation will result in more effective and cost-efficient actions/programmes towards sustainable use and conservation of local breeds in Europe

1.3 The EURECA approach

There are many different factors that help a breed to become self-sustaining or that may influence the risk status of a breed. Some factors clearly contribute to the success of particular breeds. Other factors may lead to a critical situation for the breeds. We can distinguish between genetic and non-genetic factors that can affect the endangerment status of a breed.

There are different kinds of stakeholders that may or may not improve the chances of a breed's long-term survival. Individual farmers, breeders, breed societies, breeding industries, farmer organisations, government organisations, research organisations and non-governmental organisations play a role. It has to be recognised that the initiatives of particular stakeholders and the interactions between stakeholders determine the future of the breeds.

In order to get a better understanding of the factors affecting the demographic dynamics of local/regional cattle breeds, we collected and analysed different sources of data. Fifteen breed cases across eight European countries were selected to study the history, the status, and the development of the breeds. Farmer interviews and perceptions of stakeholders and experts (National Coordinators) were used to collect new data and to analyse the breed situation within European countries.

For each breed we analysed the main strengths, weaknesses, opportunities and threats (SWOT) and in this way completed the individual breed assessments. For a selected number of breed cases we identified strategic opportunities and interacted with relevant stakeholders in order to develop or strengthen the future prospects for a breed.

In addition to the detailed assessment of 15 breed cases, we studied similarities and differences between national cryopreservation programmes, given the relevance of cryopreservation for the long- and short-term conservation of genetic diversity in local cattle breeds. Because genetic variation is vital for the survival and development of breeds, we reviewed available methodologies and software, which can be useful in helping to assess the management of genetic variation within populations.

1.4 This publication

The aim of this publication is to share experiences and information with stakeholders in European countries in order to understand the self-sustainability of a breed better and to improve strategies and policies.

In the first chapters we analyse and describe the state and trends in cattle diversity in Europe and identify common or breed/country specific factors that affect the sustainability of local breed farming. In the final chapters we examine decision-making tools for better genetic management of animal populations and for the development of policies and breed strategies in general, followed by conclusions and recommendations in the last chapter.

In Chapter 2 we describe major developments in cattle production and relevant European policies affecting cattle production. The EURECA breed cases will be put into the context of general trends. Chapter 3 describes the state of cattle diversity in Europe, including the results of a Europe-wide survey on local cattle breeds. Chapter 4 shows in much more detail the similarities and differences found in the breed cases studied. We found common factors, but more often we found differences among breeds and countries. Chapter 5 illustrates in detail the role of cryopreservation for the long- and short-term conservation of genetic diversity in cattle breeds and the different options for organising cryopreservation activities on a national level.

In Chapter 6 we show which factors contribute to breed survival from a genetic point of view and which methods and (software) tools are available for analysing the variation and

Farmer's quote:

'When concentrates become more expensive and milk prices decline, this will be an opportunity for local cattle breeds'

their utilisation and management in livestock populations. Chapter 7 shows how SWOT analysis methodology can be used to identify opportunities for individual breed strategies and how a multi-stakeholder process can contribute to strengthening breed strategies and general policies. The final Chapter 8 discusses and summarises the main conclusions and recommendations.

Overall, the EURECA project used different methods to collect data and interviewed different stakeholders in order to evaluate and understand the dynamics of local breeds better. The project is unique in its approach and will hopefully contribute to the development of successful future strategies for (other) local cattle breeds in Europe. We have found success stories and we also seem to know why things have gone wrong. We have met enthusiastic people, and governments have taken actions as have the breeding organisations that are looking at the local breeds in a new way.

References

FAO, 2007. The State of the World's Animal Genetic Resources. FAO, 2007.

Chapter 2

Trends in cattle diversity and cattle production in Europe: from popular to niche

Katriina Soini and Yvette de Haas

In this chapter:

- Description of the main factors that have accelerated displacement of local cattle breeds by mainstream breeds in Europe.
- Discussion on similarities and differences of these factors and trends between breeds and countries.
- Preview of how the future of the European local breeds may look.

2.1 Introduction

Cattle started living in close proximity to humans in Central Europe approximately 7500 years ago (Benecke, 1994) and in the Northern parts of Europe a thousand years later (Gkliasta *et al.*, 2003; Cymbron *et al.*, 2005). Since then cattle have played an important role in food production as well as in social, cultural and political development of the European societies and individual farmers. Currently, the FAO database DAD-IS lists 534 cattle breeds in European countries, of which 464 are classified as local or regional (www.fao.org/dad-is). Cattle have been used for many purposes from food production to transportation, and from draught power to social insurance. They have also been pets and ritual animals, like bulls in Spain. Today, many cattle breeds have been developed for either milk or beef production, and production is also geographically concentrated and specialised. Many European cattle breeds have disappeared (130, according to DAD-IS), or are now under threat of extinction. The key questions remain: what have been the main drivers for this trend in cattle diversity and in animal production in Europe, and what does the future hold for the local breeds?

In the following we will give a brief overview of the history of cattle production and cattle diversity development in Europe, illustrated by the selected breed cases from eight European countries examined in the EURECA project. The investigations are based on existing literature, farmer questionnaires, analysing nationally recorded data, and expert views. The breeds selected for the study represent milk, beef and dual-purpose cattle breeds. In this chapter we focus mostly on the local breeds in North, West and South-West Europe.

The map in Figure 2.1 shows us where the local breeds are located. The breed cases that are examined are:
- Brandrode Rund – Deep Red Cattle (Netherlands)
- Groninger Blaarkop – Groningen White Headed (Netherlands)
- Maas Rijn IJssel – Meuse Rhine Yssel (Netherlands)
- Ferrandaise (France)
- Villard de Lans (France)
- Reggiana (Italy)
- Modenese (Italy)
- Länsisuomenkarja, Lsk – Western Finncattle (Finland)
- Itäsuomenkarja, kyyttö – Eastern Finncattle (Finland)
- La Pie Rouge de Type Mixte – Dual-Purpose Red and White (Belgium)
- Blanc Bleu Mixte – Dual-Purpose Belgian Blue (Belgium)
- Alistana Sanabresa (Spain)
- Avileña-Negra Ibérica – Avilena-Negra Iberica (Spain)
- Polska Czerwona – Polish Red (Poland)
- Kerry cattle (Ireland)
- Estonian Native (Estland)

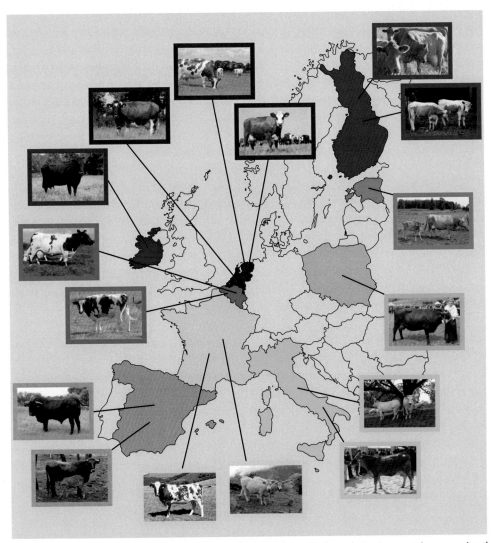

Figure 2.1. Map of Europe with location of 16 local cattle breeds in 9 countries examined within the EURECA project.

More information on each breed can be found in the boxes throughout the book.

2.2 Interest in breeds and breeding increases

A predominant view, supported by genetic research, is that European cattle breeds descend from *aurochs* domesticated in the Near East about 11,000 years ago and brought to Europe through the migration of human populations and trade (e.g. Clutton-Brock, 1999; Grigson,

1980). In the course of time, cattle became locally and regionally differentiated, as they adapted to the different local environmental conditions. The way in which people were using, feeding and maintaining cattle also affected the further development of breed characteristics, such as size, behaviour, phenotype and productivity. The selection of animals has been based on human needs and values, which have varied from place to place over time. As an indication of regional differences among cattle breeds, the name of a breed usually includes its geographical origin in addition to factors that describe the phenotype of the animals (e.g. Avileña-Negra Ibéria, Spain; Groningen White Headed, Netherlands).

In many European countries there have been various forms of initial work since the 16th century, focusing on the appearance and performance of the cattle. The work stems from an interest in exploring and comparing the characteristics and differences among breeds. In addition, the efficiency of cattle farming has become more important, especially since the 19th century as many European societies suffered from famines. On the other hand, industrialisation and urbanisation started to change food consumption habits putting a great stress on the rate of cattle production. Cattle products also became an important trading commodity (Carlson, 2001; Tervo, 2004).

Individual people imported breeding animals from abroad, and in some countries these activities were supported by the state as well. For example, in the mid 19th century the Finnish Senate provided financial support for the import of the Ayrshire breed to improve the native breeds (Lilja, 2007), and French 'Conseil Général' supported the purchase of Auvergne or Swiss bulls. On the other hand, from early on, the state also regulated cattle imports. This was necessary to prevent the spread of new diseases which might have been introduced by imported cattle, as happened in Finland in the late 19th century. In the Netherlands cattle were imported from Germany (red and white) or Denmark (black and white) after cattle disease outbreaks, like cattle fever, or after the rivers had been flooded and the cattle population was decimated.

The development of herdbooks in various parts of Europe can be seen as a sign of a growing interest in cattle performance and breeding. The first herdbooks were established in Great Britain as early as in the 18th century (Lush, 2008). Records on animals in the breeds were at first maintained by dedicated individuals, but eventually breeding associations were formed to control and maintain registration. Some of the local breeds examined in the EURECA project already had their own herdbook at the start of the 20th century (Eastern Finncattle, Finland; Ferrandaise, France; MRY, Netherlands), whereas in some other breeds the recording scheme was founded as late as 1970 (e.g. Avileña-Negra Ibéria, Spain).

Agricultural shows have acted as an important forum for the local, regional or national farming communities to promote and encourage improvements in cattle breeding in many countries. The first cattle shows were organised as early as the late 18th century. The traits of the cattle and differences between breeds (e.g. the favoured colour and other exterior

Lãnsisuomenkarja, Lsk – Western Finncattle

History

The Finnish cattle breeds belong to the distinct northern group of Fennoscandian cattle breeds. Over time the breeds have had genetic inputs from neighbouring cattle populations. About a hundred years ago the native Finncattle was divided into three breeds – Eastern, Northern and Western Finncattle. The Western Finncattle (WFC) has long been the most common breed in the favourable agricultural regions. Now they are found in other regions as well. WTC animals are beige-brown, with some occasional white markings or spots. Practically all present-day animals are polled.

Breeding, conservation and promotion

The Finncattle population started declining when the Ayrshire and Holstein were introduced ('50-'60). New attention was focused on Finncattle breeds in the 1980's. When Finland joined the European Union in 1994, the subsidy programme was set up to support the farms keeping native Finnish breeds. Now there are some 3,000 WFC individuals linked with 1,700 herds in the animal register, with two thirds of the cows belonging to the milk recording scheme. The breeding programme aims at improving the milk production and functional traits. The semen storage contains over 260,000 doses from 160 bulls, with about 40 born before 1980.

SWOT

S: High dry matter content of milk; high efficiency in low-input conditions; middle-sized animal fits nicely into old cow sheds.

W: Low profitability; small population size for efficient selection schemes.

O: Suitable for branded cheese and other products.

T: Declining number of farms and missing partnership.

traits) were discussed and debated at the fairs, and later reported in the press. Local breeds that participated in these shows were compared with other local breeds as well as with imported cattle breeds. The French Ferrandaise breed had the place of honour at the regional agricultural show, in Clermont-Ferrand in 1863, and the Italian Reggiana breed was present at the Vienna Expo in 1873.

Farmer's quote:

'Local cattle is a living inheritance, highly valued and respected; old germplasm which should not get lost'

In Spain, animals of local breeds are found in old documents from the 'Exhibición General' in Madrid (1857), where local animals were compared with imported animals mostly Brown Swiss and Friesian. In addition, documents of exhibitions in the 19th century and early 20th century refer to 'vaca del pais' (i.e. 'cow from the country') as opposed to the specialised animal brought from outside (M.A.P.A., 1887).

In the late 19th century local breeds also had a political significance outside agriculture. In Finland the awareness of national identity increased in the late 19th century, and the real and pure Finncattle, unlike their foreign relatives, were considered as part of the national mindset and became a subject for some artists' work. A sense of patriotism was cited as a reason for breeding them (Lilja, 2007).

2.3 Local breeds in and between the wars

In the late 19th and early 20th century there was more systematic work on improving cattle to meet the growing demand for food. At that time in several countries there were institutional arrangements that supported cattle breeding activities, e.g. the establishment of agricultural extensions and breeding organisations. Investments were made in research on animal production and breeding. It was realised that efficiency could be increased by improving the feeding and breeding of the animals. In addition, the trade of agricultural products between countries increased, for example Dutch Friesian cattle were sold to the US because of their milk production. In Finland the income from exported butter was used to buy foreign cereal products. Governments started to regulate cattle breeding and decided both in Italy and the Netherlands that only sires registered in an official herdbook were allowed to reproduce. In both countries this implied that bulls of many local breeds were no longer official breeding bulls, because for these breeds no official herd book was established at that time.

The influential development in agriculture was interrupted by the World Wars. The local breeds suffered from the war periods in many ways: a lot of cows were killed due to the war activities. These breeds were slaughtered by the invading troops (Finland) or by hungry citizens (France) for food. In addition, cattle breeds were moved from their original regions to new places with the people that were evacuated during the war (e.g. Eastern and Northern Finncattle), and cattle were even exported to other countries (Villard de Lans, France).

Right after the Second World War the first task in many countries was to prevent famine among the human populations and safeguard the self-sufficient nature of the national food production system. Local cattle breeds, that were often dual or even multi-purpose, were well-suited to this purpose, because they provided both milk and meat and therefore generated a respectable income for small farms (e.g. Dual-Purpose Red and White and Belgian Blue in Belgium; Eastern and Western Finncattle, Finland; and MRY, Netherlands).

2.4 Modernisation of agricultural production

Soon after the Second World War there was an urgent need to increase production, and consequently new goals were set for agricultural production in order to improve the efficiency and productivity of farming. Many technological innovations contributed to the move towards more efficient production. The intensification of agriculture required investments and there was a need to replace local breeds with high-input high-output breeds to improve the overall efficiency of cattle production.

The introduction of tractors in the late 1950's resulted in a decline in the local breeds which had been used for draught power until that time (e.g. Ferrandaise, France; Alistana-Sanabresa and Avileña-Negra Ibérica, Spain). In Spain, for example, the triple-purpose animal had suited the needs of the majority of the population: the draught animals had been used to cultivate cereal, the young calves were sold to regional markets and the milk was mostly used for family consumption. The 1960's were a turning point: draught animals were replaced by tractors and the demand for animal protein increased. These changes severely affected the Spanish triple-purpose breeds, which have always been closely linked to cereal production in Castilla (local breeds were used for draught power purposes, both for collecting the crops from the field and for transporting them to Madrid). Once tractors had arrived, the draught power of cattle was no longer needed and the population size of the local cattle breeds declined rapidly. The increasing demand for animal protein required specialised beef breeds with higher beef production, instead of triple-purpose animals with lower beef production (Fontana, 1981).

Milking machines were introduced on a large scale in the 1960's and 1970's. They had been designed primarily for the udders of commercial dairy breeds (Holstein Friesians), which had been selected much earlier for correct shape and teat placement. The conformation of the udder of some local breeds (e.g. Groningen White Headed, Netherlands; Reggiana, Italy) was not satisfactory, and caused difficulties when attaching the milking machine equipment. For this reason, the local breeds were no longer favoured by many farmers.

The development of artificial insemination (AI) and the use of frozen semen enhanced the extensive dissemination of mainstream breeds (such as Holstein Friesians or French beef breeds) and also initiated large-scale crossbreeding with local cattle breeds. This accelerated the introgression of new genes in many local cattle populations. Importing semen from other countries for crossing with the native breeds occurred in the 1950's in Italy (US Holstein), and in Finland in the 1960's, when Finncattle were systematically crossed with Friesian cattle. In Belgium, Canadian Holsteins were imported in the 1970's and crossbreeding in the Netherlands also started in the 1970's when Dutch Friesians, MRY and other native breeds were crossed with Holstein Friesians. In Spain, Avileña-Negra Ibérica, which is a typical beef breed, was mated with 'Vaca Holandesas' (Dutch Friesian) to improve dairy aptitude. This shows that AI also played a role in speeding up the specialisation of breeds in either milk or beef production.

Improved veterinary skills in caesarean techniques paved the way to selecting for double-muscled Belgian Blue cattle notorious for difficult calving. The Belgian Blue cattle were attractive because the demand for good meat quality increased between 1960 and 1970. Therefore, the Dual-Purpose Belgian Blue started to become obsolete, and many breeders started to select double-muscled animals.

In the old days, cattle were simply fed on natural pastures and feed for cattle was collected from marginal lands. Increased cattle production was possible because arable land for crop production was released for intensive feed production and high quality feed was imported from other continents. Chemical fertilisation made it possible to increase the amount of harvested yield per hectare. At the same time feed storage systems were improved and feed could be harvested within a short period and fed many months later. As a result of this development, feed quality was much more consistent throughout the year, which is especially important for the high-yielding breeds (Holstein Friesians) which are much more demanding than the local cattle breeds (e.g. Irish Kerry Cattle; Dutch Deep Red Cattle). All these changes in farming enhanced the use of high-input high-output Holstein Friesians in dairy production.

The modernisation of agriculture was accelerated by national agricultural policy activities. In addition, the EU Common Agricultural Policy (CAP) was first introduced in 1962, and still plays a very central role in European agricultural and rural development today. The goal of the first CAP was simply to increase agricultural productivity by stimulating technical improvement and efficiency of agricultural production, ensuring proper incomes for the agricultural population, stabilising markets, ensuring food availability, and providing consumers with food at reasonable prices (Article 39).

2.5 Extending food markets

Together with agricultural improvement, socio-economic and cultural developments in European societies have affected European cattle production and cattle breed diversity.

In some European countries industrialisation had already started at the beginning of 19th century and was accompanied by rapid urbanisation. It led to a rapid growth in the human population, new division of labour, increased wealth, a widening and diversification of food markets with the cost of decreased self-sufficiency in production, and to alienation from the rural way of life and nature.

The intensification of farming and the aforementioned technological developments made it possible to introduce new products for the consumers. For example, yoghurt and low-fat milk products in many countries became popular among the rising urban population. Moreover, the development of household techniques, like freezers and fridges, increased the consumption of dairy and meat products and made more efficient or specialised production very profitable. In Spain, for example, beef consumption per person went up from 5.9 kg per person per year in 1960 to 14.1 kg in 1975, but decreased again to 7.7 kg in 2004 mainly due to the relative increase in beef prices compared with meat from other animals, and to the preference for 'easy to prepare'-meat that is low in cholesterol. These changes have also been affected by the cattle feed safety scandals (BSE, dioxin). During the last decades of the 20th century, national food cultures and food demands have changed dramatically and the amount of imported food has increased.

2.6 Decline of local cattle breeds

Most of these developments have been harmful to the local breeds. The desire for increased productivity and profitability has led to specialisation in either milk or beef production and to a decline in the use of dual or multi-purpose cows that was emphasised earlier between the 1950's and 1980's. The focus on intensification and specialisation of animal production was also reflected in the breeding goals. The advantages and disadvantages of various local and commercial breeds were debated at many forums, and in most countries the local cattle breeds ended up being the losers in these breed-confrontations. Local breeds were undervalued and thought to be both low-producing and old-fashioned. The farmers themselves felt pressure to switch from the local breeds to the modern and efficient dairy or beef breeds.

The decline in the number of heads of the local breeds took place in various countries at varying speeds and in slightly different ways. In Spain, for example, 74% of registered cows in 1955 used to be of a local cattle breed. In 1986, that percentage had fallen to 26%. Between 1950 and 1980, the number of cows from local cattle breeds went down from almost 1.3 million to 700 thousand. Conversely, between 1950 and 1980 the numbers of Friesian cows

Farmer's quote:

'The schoolchildren come to see the different cattle breeds. With local breeds you have something special and exclusive'

went up from 338 to 1,375 thousands heads. Similarly, crossbred cows and cows from specialised breeds duplicated their census (García Dory, 1986). In Finland, as another example, the population of Western and Eastern Finncattle declined dramatically from approximately 500,000 in 1950 to only few thousand today (FABA; see also Figure 2.2). In the Netherlands, within 30 years the proportion of local breeds in total dairy production decreased from almost 100% to less than 5% (Buiteveld *et al.*, 2009). During this period, complete extinction of local cattle breeds was avoided by the prompt activities of some individual people and farmers, who were committed to keeping the local breeds. The breeds have also survived in remote rural areas.

2.7 New policy turns

Agricultural policies practised in the 1960's and 1970's aimed at intensifying agricultural production and accompanied by agricultural modernisation had a dramatic impact on agricultural environment and rural society. They resulted directly or indirectly in (1) product surpluses, (2) growing export subsidies and (3) loss of biodiversity in the agricultural

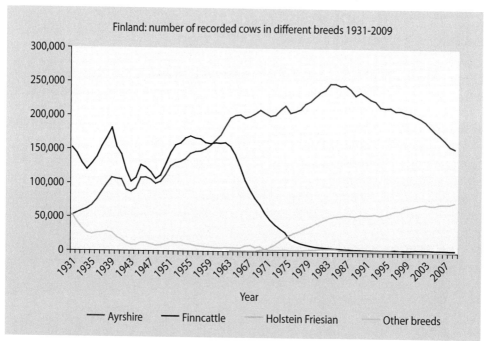

Figure 2.2. Development of the populations of the Finncattle versus imported Ayrshire and Holstein Friesian cattle in the recorded herds. The census is provided by FABA Finland, which provides services ranging from artificial insemination and embryo transfer to genetic evaluations and breeding advice.

environment including a reduced range of local cattle breeds. Many farmers, like other rural dwellers, moved from the countryside to urban areas to work in industry or services. Those who stayed in cattle production intensified their farming and specialised in either dairy or beef production, with a breed that best suited that purpose.

As a result of these new trends, new regulations were needed. The milk quota system was introduced in 1984 in Europe to stop the over-production of milk. In 1992, agri-environmental measures (Council Regulation (EEC) No 2078/92) were introduced as part of the Common Agricultural Policy to support and encourage farmers to protect and manage the environment on their farmland better. This programme also included a measure for keeping indigenous rare/local breeds. A survey was carried out in Europe by Small and Hosking (2010) to determine how Farm Animal Genetic Resources are supported in both EU member and non-member states. They received responses from 18 countries, and 11 of them used the national allocation from the EU Rural Development Programme to support the conservation of Farm Animal Genetic Resources. All but two paid on a Livestock Unit basis. The intended aim of this measure is to benefit genetic diversity, but there is also expected to be a positive impact on the landscape.

There has also been a focus at the international level on the loss of biodiversity in agriculture. The Convention of Biodiversity (CBD) was negotiated within the framework of the UN Environment Programme in 1992, and was ratified by the European Union and other European states. The three objectives of the CBD are: (1) conservation of biological diversity, (2) a sustainable use of components of biological diversity, and (3) fair and equitable sharing of the benefits arising from the utilisation of genetic resources. Although not directly stated in the CBD, these objectives also include animal genetic resources. The nations that have ratified the Convention have accepted responsibility for the utilisation and conservation of the national farm animal genetic resources. Furthermore in 2007, member states adopted the FAO Global Plan of Action for Animal Genetic Resources, which is an important milestone and trigger for European countries to design and further develop a national strategy and action plans for farm animal genetic resources.

2.8 21st century: new opportunities for local breeds?

After a few less glorious decades, the new millennium seems to be offering many opportunities for local breeds. In the late 1990's new food trends and initiatives like 'functional food', 'local food', and 'slow food' provided an opening for breed-specific products of local breeds, as they could easily be associated with all these trends. Food is considered not only as a means of nutrition, but increasingly as a way of living and part of someone's identity (e.g. personality, social status): 'You are what you eat'. Therefore, there are growing markets for niche products, or products differentiated along with various characteristics, such as the method and place of production and/or cultural traditions. It is not only the image of the product, but also the gastronomic characteristics and quality that determine consumption patterns. Milk from many local cattle breeds seems to be very suitable for cheese-making due to its high dry-matter content and protein composition (Eastern Finncattle; Reggiana, Italy; Groningen White Headed, NL) and the beef of some local breeds is appreciated for its cooking quality.

There are already brand-products like (1) Parmigiano Reggiano cheese made with Modenese or Reggiana cattle milk only, and (2) branded Modenese meat, under the Consortium 'Valorizzazione prodotti bovini di razza Bianca Valpadana Modenese' in Italy. The Slow Food-movement has also created a 'presidium' for some breeds and their products. In France, a special trademark 'La Ferrandaise' was created in 1999 and a 'Ferrandaise' restaurant was opened. In Spain the first label linked to beef 'Carne de Avileña' and the 'Consejo Regulador de Carne de Avileña' was created in 1988 to promote and control the meat sold under this specific brand. Later in 1993 the name was changed to 'Carne de Avila' to adapt to the EU legislation. In 1999, the association APTA (Aliste Beef Promotion Association) in Spain created a Guarantee Label, 'Ternera de Aliste'. In the Netherlands, some farmers make branded cheese with milk from either Groningen White Headed or Deep Red Cattle (called 'Leidse Sleutelkaas' and 'Brandrood Kaas', respectively). In Finland, very recently, new collection of cheeses called 'Armas-family' made from the milk of Eastern Finncattle was introduced on the initiative of a Finnish chef, and active co-operation between some Finnish restaurants and local cattle farmers has emerged. There is also an increasing demand for the skins of the Eastern Finncattle cows.

In addition to the new food consumption trends, the policy reforms that have been made in the 21st century offer many new opportunities for the local breeds, but also some drawbacks and threats. According to the CAP reform in 2003 the majority of subsidies will be paid independently of the volume of production. It is assumed that removing the link between subsidies and production will make farmers more competitive and market orientated, whilst providing the necessary income stability. In addition, mitigating the problem of over-production, subsidies on livestock products and animals have been restructured and there are plans to eliminate the milk quota system, which, however, may put an greater higher pressure on production efficiency.

The European Rural Development Policy has accompanied and complemented the Common Agricultural Policy since 2000. The Rural Development Programme for 2007-2013 aims not only to improve the competitiveness of the agricultural and forestry sector and the state of the rural environment, but also to improve the quality of life in rural areas and to encourage diversification of the rural economy (EC 1698/2005). The programme contains measures on diversification towards non-agricultural activities, support for the establishment and development of micro-business, promotion of tourism, and the protection, development and management of natural heritage. All these measures are contributing to sustainable economic development. It can be argued that conservation and preservation of local breeds fits well with all these aims: the breeds are able to contribute to the efficiency of animal production and safeguard the possibilities for genetic diversification in the future. They are also contributing to the landscape and biodiversity management of rural areas. Breed-specific

Brandrode Rund – Deep Red Cattle

History

The Dutch Deep Red cattle used to be part of the Meuse-Rhine-Yssel (MRY) breed. After the Second World War, the colour pattern of MRY changed to more white and light red. Some farmers continued breeding with their own bulls and so their more 'deep red' cattle became genetically isolated from the more white population. Nowadays the Deep Red cattle has become an official breed. Deep Red cattle are extremely suitable as free roaming cows in nature reserves because of their robustness. The population shows an increasing trend.

Breeding, conservation and promotion

There is an active breed society, called 'Het Brandrode Rund'. In 2007, 454 pure-bred Deep Red cows and 66 pure-bred Deep Red bulls were registered in the herdbook. The breed society is very active in organising the breeding programme, in selecting bulls, and in promoting the breeds. Inbreeding is a major point of attention. In 2008, semen from 12 Deep Red bulls as well as 14 embryos were frozen and stored in the Dutch Gene bank (CGN).

SWOT

S: The foundation 'Het Brandrode Rund' actively promotes the breed, and 'Brandrode Rund' is also used as a brand name.

W: The maintenance of the breed depends largely on volunteers and hobby farmers.

O: An important opportunity is to make money with special breed-related products. This will improve the profitability and also the brand awareness of the breed.

T: The most significant potential threat for the Deep Red cattle is the small population size, without a clear breeding goal.

products of local breeds may contribute to the rural economy and increase the quality of life in both rural and urban areas. The breeds may also have a wider impact on rural society, as they might serve as a source of income and a possibility to stay in the sparsely populated rural areas (Soini, 2007; Soini and Partanen, 2009).

In addition to the agricultural and rural policies, EU has designed new legislation for veterinary and food safety, mainly to cope with newly-discovered health hazards, like BSE and dioxin contamination. Therefore, there is now a considerable amount of food safety legislation, on animals and animal products within the EU, but also in third countries on the exports from these countries to the EU (http://europa.eu/legislation_summaries/food_safety/index_ en.htm). On the other hand, maintenance of local breeds and development of local food products may be hindered by this kind of legislation, because big investments are needed by a farmer to comply with the regulations. For the small-scale farmer with a few local breed cows, the investments are particularly high for committing to a specific on-farm product (e.g. cheese). Moreover, the EU zootechnical legislation is regulating free trade in breeding animals and their genetic material (http://ec.europa.eu/food/animal/zootechnics/legislation_en.htm). It also lays down rules for entering animals into herdbooks and for the recording of performance data and estimation of breeding values and acceptance for breeding purposes.

All in all, the recent policy developments include a number of options for supporting conservation and the use of local breeds, ranging from direct subsidies to reduce the gap in profitability per animal between the local and mainstream breed, to subsidies for (1) non-productive components like environmental services, cultural heritage and (2) supporting institutions, breed interest groups or breed societies. These subsidies can be implemented in rural development programmes on a national level. It has to be added that national governments often have higher priorities than local breeds.

As a result of developments in agricultural, socio-economic, cultural and policy fields, the status of the local cattle breeds has changed rapidly during the 20th century from a major breed to a marginal one. This development has been taken place in all European countries, although at a different pace and in different ways. However, many common turning points can be identified (Figure 2.3). As a conclusion it can be argued that although there are still many threats to the local breeds in economic, political and social terms, there are also mechanisms that may promote the self-sustainability of local breeds, or at least prevent their extinction.

2.9 Conclusions

As a result of developments in agricultural, socio-economic, cultural and rural and environmental policy fields, the status of the local cattle breeds has changed rapidly during the 20th century from a major breed to a marginal one. Examples of such developments are (1) the increased interest in breeds and breeding, accompanied with the establishment of

Society	Agricultural society		Industrial society				Post-industrial consumer society			Towards more sustainable society?
Year	1800	1900	1950	1960	1970	1980	1990	2000	2010	2020
		Great famine 1866-1868	From self-sufficient households to food markets and trade	Yoghurts and low-fat milk products introduced	Arrival of international food cultures	Food crises; discussion about quality of food starts		Movement on local food emerges	Products of local breeds available in the markets and restaurants	
		Importance of milk production increases	Agricultural research and extension organised	Dairy co-operatives, tractors, milking machines	Overproduction, payment for slaughtering cows	Milk quota system introduced 1982		Membership of EU 1995 & CAP	Rapid decrease in number of farms	
	State imports breeds	Imports banned due to animal diseases	Productivity requirement was included in herdbook in 1912	Discussion about differences between Finncattle breeds	Crossbreeding of Finncattle with Holstein-Friesian and Ayshire	Establishment of cryobank	First Eastern Finncattle embryos stored	Separation of Eastern, Northen and Western Finncattle in animal register		
		Association & herdbook for Eastern Finncattle in 1898	Herdbook for Western Finncattle in 1906	Joint association for Finncattle breeds	Attention to critical situation of Finncattle		Finncattle housed in three prisons for conservation	Agri-environmental subsidies for keeping Finncattle 1995 ->	Eastern Finncattle receives attention in the media	

Figure 2.3. Main turning points in developments in agricultural, socio-economic, cultural and policy sectors affecting Finncattle (compiled from Lilja, 2007).

herdbooks and agricultural shows, and (2) the modernisation of agricultural production with technological innovations, industrialisation and urbanisation. These developments took place in all European countries, although at a different pace and in different ways. However, many common turning points can be identified.

Farmer's quote:

'Local breeds are like diesel cars. Slow to start, run on cheap fuel and keep going for a long time'

There are still many threats for the local breeds in economic, political and social terms. Younger farmers do not necessarily understand the value of local breeds, and they often see local cattle breeds as low-yield and old-fashioned. If no-one continues to farm local cattle, there is a risk that this way of living will disappear. The global agricultural and animal market can be considered as a threat as well. On the other hand, there are also initiatives that promote local breeds. They contribute to more self-sustaining local breeds, and at least prevent their extinction. The introduction of breed-specific products is one example of an efficient and effective initiative that promotes local breeds.

As regards the future of local breeds in Europe, there are many ongoing economic, social and cultural processes, which open up new possibilities, such as increased environmental awareness, more sustainable ways of living and individual consumption habits, making livestock production systems more sustainable, (re-)discovering local/regional/national identity, or new functions and roles of farm animals in society (e.g. nature management, care farming, education services).

References

Benecke, N., 1994. Archäozoologische Studien zur Entwicklung der Haustierhaltung in Mitteleuropa und Südskandinavien von den Anfängen bis zum ausgehenden Mittelalter. Schriften zur Ur- und Frühgeschichte. Akademie Verlag Band 46, Akademie Verlag Berlin, Germany.

Buiteveld, J., Van Veller, M.G.P., Hiemstra, S.J., Ten Brink, B. and Tekelenburg, T., 2009. An exploration of monitoring and modelling agrobiodiversity: From indicator development towards modelling biodiversity in Agricultural systems on the sub-specific level. CGN-report 2009/13 45 pp.

Carlson, L.W., 2001. Cattle. An Informal Social History. Chicago, Ivan R. Dee.

Clutton-Brock, J., 1999. A Natural History of Domesticated Mammals (Cambridge Univ. Press, Cambridge, U.K.), 2nd Ed.

Cymbron, T., Freeman, A.R., Malheiro, M.I., Vigne J.-D. and Bradley, D.G., 2005. Microsatellite diversity suggests different histories for Mediterranean and Northern European cattle populations. Proc. R. Soc. B 272: 1837-1843.

Fontana, J., 1981. Cambios económicos y actitudes políticas en la España del siglo XIX. Ed. Ariel, Barcelona.

García Dory, M.A., 1986. Las razas bovinas autóctonas de España están en peligro de extinción. Quercus 23: 14-21.

Gkliasta, M., Russell, T., Shennan, S. and Steele, J., 2003. Neolithic transition in Europe: the radiocarbon record revisited. Antiquity 77: 45-62.

Grigson, C., 1980. The craniology and relationships of four species of Bos. 5. *Bos indicus L.* J. Archaeol. Sci. 7: 3-32.

Lilja, T., 2007. Suomalaisten maatiaislehmien vaiheet omavaraistaloudesta 2000-luvulle. In: Karja, M. and Lilja, T. (eds.) Alkuperäisrotujen säilyttämisen taloudelliset, sosiaaliset ja kulttuuriset lähtökohdat. Agrifood Research Finland, Publications 106, Helsinki.

Lush, J.L., 2008. Animal Breending Plans. Fourth edition. (First edition 1943). Orchard Press.

M.A.P.A. 1887. La crísis agrícola y pecuaria. Tomo III, Sucesores de Ribadeneyra. Madrid.

Small, R. and Hosking, J., 2010. Rural Development Programme Funding for Farm Animal Genetic Resources: A Questionnaire Survey. Report for the National Standing Committee on Animal Genetic Resources, UK.

Soini, K., 2007. Maatiaseläinten monet arvot. In: Karja, M. and Lilja, T. (eds.) Alkuperäisrotujen säilyttämisen taloudelliset, sosiaaliset ja kulttuuriset lähtökohdat. Agrifood Research Finland, Publications 106, Helsinki. 17-39.

Soini, K. and Partanen, U., 2009. The Golden Stock. In. Granberg, L., Soini, K. and Kantanen, J. Sakha Ynaga: cattle of the Yakuts. Suomalaisen tiedeakatemian toimituksia. Humaniora 355: 169-188.

Tervo, M., 2004. Lehmä (A Cow). Atena Publication, Jyväskylä.

Chapter 3

State of local cattle breeds in Europe

Delphine Duclos and Sipke Joost Hiemstra

In this chapter:

- Overview of the history, diversity, classification and risk status of local cattle breeds in Europe.
- Results from a Europe wide survey on local cattle breeds among National Coordinators.

3.1 Introduction

According to the State of the World's Animal Genetic Resources (FAO, 2007), about 20% of all breeds in the world are considered to be 'at risk' (number of breeding females fewer than or equal to 1000). For local cattle breeds in Europe, this figure is even more alarming. Based on the information in the European and global databases (e.g. EFABIS (http://efabis.tzv.fal.de/) and DAD-IS (www.fao.org/dad-is/), respectively) 40-50% of the local cattle breeds in Europe can be considered to be 'at risk'. The percentage of local breeds at risk increases dramatically when the EU-threshold for support of endangered cattle breeds (EC, 2004; EC, 2005) is used, which is below 7,500 cows.

In European and global databases there is a distinction between breeds that appear in only one country and those that live in more than one country. The first are referred to as 'local' breeds, and the latter are referred to as 'transboundary' breeds. Within the transboundary breed category, a further distinction is made between 'regional' transboundary breeds: those that appear in more than one country, but within a single region, and 'international' transboundary breeds: those that have spread across countries and to more than one region, such as Holstein Friesian (FAO, 2007). Currently, the majority of cattle breeds in Europe can be categorised as local breeds. Those breeds are recognised to have important economic, environmental, social, cultural, historical and genetic values. In this book we use the term 'local breeds' including both local and regional transboundary breeds.

3.1.1 History of local cattle breeds in Europe

From the EFABIS and DAD-IS databases we count more than 400 local cattle breeds in Europe. Several molecular genetic analyses have been done to better understand the origin and genetic differentiation of European breeds. Archaeological evidence indicates that there were two main routes for cattle to enter Europe from domestication centres: (1) the Danubian route across the central European plains, and (2) the Mediterranean route along the coast of the Mediterranean Sea. The hypothesis has been confirmed by microsatellite analysis (Cymbron *et al.*, 2005). Another molecular study has found that two main groups of breeds can be distinguished in Europe: (1) Podolian breeds covering for example many Italian and Hungarian breeds, and (2) other cattle breeds (Negrini *et al.*, 2007).

3.1.2 Between- and within-breed genetic diversity

Molecular analyses are also used to measure diversity both between and within breeds. The formation of breeds is a redistribution of a common base variation. Numerous breeds are carrying along most of the original variation while the ones that have gone through bottlenecks or experienced long periods of inbreeding have less variation left. A close relationship between breeds can be deduced from similarities in the exterior traits (shape, colours, shape of horns, etc.), whereas the estimation of more distant relationships would require an assessment of similarities and differences in the genome. The assessment can be performed with neutral markers or with those that have a known connection to the variation in adaptive or economically important traits (e.g. Toro and Caballero, 2005; Toro *et al.*, 2009). The

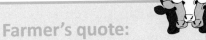

Farmer's quote:

'Cows of local cattle breeds are perfect for inexperienced farmers; they are easy to manage, without a lot of hoo-ha'

neutral markers reflect the overall variation and therefore indicate the magnitude of breed differentiation and potential variation in traits not yet selected.

Some studies identified breeds with particularly huge within-breed diversity: for example, Busă breeds located in the region of the Balkan (Macedonia, Albania, Bosnia-Herzegovina, Kosovo breeds) are found to show the highest within-breed diversity, whereas alpine or north-western European breeds show very low within-breed diversity. The decrease in diversity with increasing distance from the domestication centre is evident in cattle (Medugorac *et al.*, 2009).

Molecular data are sometimes used to indicate conservation priorities, but it is important not to forget that methods still need to be improved (European Cattle Genetic Diversity Consortium, 2006) and that local tradition and history (Gandini and Villa, 2003) as well as the environmental role are also essential factors when estimating the value of a breed.

3.1.3 Assessment of endangerment status

The risk status or state of endangerment of breeds is estimated based on the number of animals belonging to the breed in question and indicates whether (urgent) action is needed. Gandini *et al.* (2004) defined 'degree of endangerment' as 'a measure of the likelihood that, under current circumstances and expectations, the breed will become extinct'. Degrees of risk are difficult to assess accurately and both demographic and genetic factors must also be incorporated. Clearly, current population size is an important factor in determining risk status, but does not give the whole picture. In addition to the state of the breed population at a certain point in time, the dynamics of a breed population is even more relevant. Several criteria have been proposed to measure the state of endangerment. In this chapter we use both FAO and EU systems.

3.1.4 Towards better strategies and policies to support local breeds

In line with international obligations (CBD, 1992), and after the adoption of the FAO Global Plan of Action on Animal Genetic Resources (FAO, 2007), European countries are developing strategies for the conservation and sustainable use of Animal Genetic Resources, including local cattle breeds. One important question in a dynamic and complex world is how to influence the success of local breeds positively and how to make the breeds self-sustaining through new policies and strategies. The aim of the EU co-funded project EURECA (www. regionalcattlebreeds.eu) was to get a better understanding of sustainability in farming local breeds and the factors affecting it in Europe, which may help in defining policies and strategies.

3.2 Europe-wide local cattle breed survey among National Coordinators

In addition to existing data in the European/global breed databases, EURECA aimed to expand information on local cattle breeds, in particular on some aspects that are not considered by these databases (e.g. economic role, involvement of stakeholders etc.). We carried out a Europe-wide survey to collect information from the network of FAO National Coordinators for Animal Genetic Resources (NCs) in Europe about the state of local cattle breeds in their countries. The aim was to gain a better understanding of the state of local cattle breeds on the basis of a limited number of parameters. In total, 32 NCs were asked to fill in the questionnaire for all local cattle breeds in their country with fewer than 7,500 females (EU criteria for endangerment). From a total of 173 breeds, 108 breed questionnaires from 24 countries were returned by NCs (see Table 3.1).

Concerning the population size of breeds in the returned questionnaires, 32% of the breeds had between 1000 and 7,500 females, 55% between 100 and 1000 females, and 13% fewer than 100 females. This indicates that 68% of the breeds in the returned questionnaires are 'at risk' according to the FAO criteria (population size smaller than 1000). A large majority of the breeds reported in the survey show an increasing or at least a stable population size. In fact, only 13 breeds out of 108 were indicated as decreasing.

3.2.1 Type of farmers keeping local cattle breeds

Survey results showed that most local cattle breeds (75%) are kept in farms smaller or much smaller than average farm sizes in the respective country (Figure 3.1). In the context of the general trend towards larger farms, this may constitute a serious threat for local breeds. The percentage is even higher (92%) when we consider only the breeds with decreasing demographic trends (13 breeds).

Table 3.1. Number of breeds surveyed per country, according to number of breeding females.

Country	Number of breeds surveyed			
	No. females <100	No. females 100-1000	No. females 1000-7,500	Total
Austria	1	2	3	6
Belgium	-	-	1	1
Croatia	-	3	-	3
Cyprus	-	1	-	1
Czech Republic	1	-	-	1
Denmark	2	2	-	4
Germany	-	3	-	3
Greece	-	1	1	2
Estonia	-	1	-	1
Finland	-	2	1	3
France	1	9	6	16
Hungary	-	-	1	1
Ireland	-	1	-	1
Italy	5	8	7	20
Montenegro	-	2	-	2
Netherlands	1	4	3	8
Norway	-	6	-	6
Poland	-	1	3	4
Portugal	-	-	1	1
Serbia	-	2	-	2
Slovakia	-	-	1	1
Sweden	-	-	1	1
Spain	2	10	6	18
Switzerland	-	2	1	3
Total	13	59	36	108

Concerning the family incomes, NCs or other national experts were asked to estimate the percentage of the total family income directly linked to the local breed. This estimation was completed for less than 25% of the questionnaires. For 33% of the reported breeds the family income related to the local breed production was estimated to be lower than 25%. This low percentage can be explained by the fact that many breeders are hobby or part-time farmers and, in this case, the income associated with the local breed is not the major source of income. For many local breeds these kinds of farmers are a majority, and the average income per farm is usually quite low. On the contrary, the local breeds where the family income linked

Ferrandaise

History

The French Ferrandaise breed is a cattle breed originating from the Puy de Dôme 'département' in the 'Massif Central' (mountain range in the centre of France), not far from the city of Clermont Ferrand. Traditionally it is a triple-purpose breed (milk, beef, work). Calves were very much appreciated for veal. Milk from Ferrandaise cows was the source of typical local cheeses like 'Fourme d'Ambert', 'Fourme de Montbrison', 'Fourme de Rochefort', and 'Bleu de Laqueuille'. Today, it is not possible to associate Ferrandaise breed with a specific product or breeding system. A little more than half of the animals are bred in suckling herds. The other animals are bred in dairy herds. The breed is appreciated for its versatility both in suckling and milking systems. The total number is increasing.

Breeding, conservation and promotion

Bulls of different origins have been found and collected since 1979. Today, 29 bulls are available for artificial insemination with a good genetic diversity which prevents inbreeding (total inbreeding of females is 2%). Since 1980 a register of all Ferrandaise animals has been recognised as the official herdbook and is held by the 'Institut de l'Elevage'. Every year all breeders are contacted or visited. The female population in 2008 was around 1000.

SWOT

S: The good functional traits are recognised and the genetic variability is well preserved, particularly through the AI bulls available.

W: There is a shortage of references on the breed as well as a product directly linked to the breed.

O: The national and regional interest in the breed is strong.

T: The dependence on continuity of support from Institut de l'Elevage.

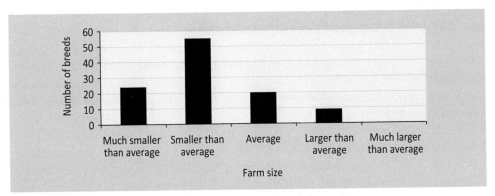

Figure 3.1. Number of breeds according to their average farm size.

to the breed was estimated to be higher than 75% are those breeds with a high percentage of full-time breeders (>75% in most cases).

3.2.2 Breeding goals

A common characteristic of local cattle breeds is that they are often not highly selected and specialised in one type of production like the mainstream breeds. This situation is also confirmed in the survey since approximately 50% of the breeds were classified as dual-purpose (Figure 3.2). Only 17 out of the 108 breeds were classified as (specialised) dairy breeds. In each category of breeds (milk, beef or dual-purpose), some breeds show a decreasing trend, but for the majority of the breeds the trend in the population size is stable or increasing (Figure 3.2).

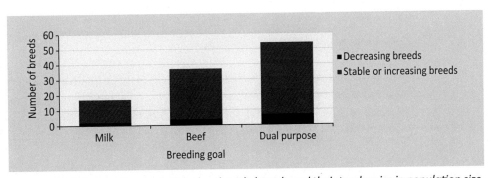

Figure 3.2. Main breeding goal of the local cattle breeds and their tendencies in population size.

3.2.3 Herdbook registration and performance recording

Registration of breeding animals is a prerequisite for a genetic management of the population (see Chapter 6). The survey showed that among the 108 breeds surveyed only four do not benefit from registration of breeding animals. NCs indicated that 75% of the breeds have more than 60% of breeding females registered in an official herdbook.

However, only a few farmers and animals participate in regular milk or beef recording (Table 3.2). Almost half of the breeds in this survey do not have any animal performance recorded. Among the 14 breeds with more than 80% of all cows milk recorded we found 5 dairy breeds and 9 dual-purpose breeds. Performance recording can be particularly expensive in breeds where animals are scattered among small herds located in difficult areas. However, basic knowledge of performances and aptitude is essential for breed development.

Table 3.2. Number of breeds according to their percentage of females registered in a herdbook and performance recorded.

Tasks	Percentages of females registered or recorded					
	0%	<20%	20-40%	40-60%	60-80%	>80%
Registered in a herdbook	4	4	10	8	17	65
Milk recorded	61	15	8	4	6	14
Beef recorded	62	14	9	7	7	9

3.2.4 Artificial insemination and cryopreservation

Artificial insemination (AI) in local breeds is not used as much as in mainstream breeds: for almost 50% of the breeds, less than 20% of cows are artificially inseminated according to the estimations of NCs (Figure 3.3). In dairy cattle, the percentage of AI is higher; 40% of dairy breeds have more than 60% of females inseminated.

For 93 out of 108 breeds, there is cryopreserved stock of semen. But the total amount of cryopreserved semen available (in genetic reserve or for routine purpose) differs greatly among breeds (Table 3.3). Information on number of bulls and amount of semen stocked (routine use or cryopreserve) was available for almost 75% of the surveyed breeds.

Cryopreservation of embryos is rare compared to cryopreservation of semen (only 26 breeds cryopreserve embryos), and the number of embryos cryopreserved is low (<50 embryos for

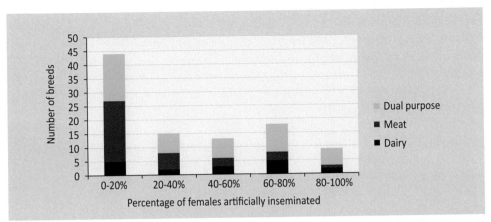

Figure 3.3. Number of breeds according to their number of females artificially inseminated.

Table 3.3. Number of breeds according to their number of semen straws available for routine AI use or in cryoreserve (information available for 84 breeds).

Number of straws	Number of breeds
<1000	8
≥1000 and <10,000	27
≥10,000 and <50,000	25
≥50,000	24
Total	84

17 out of the 26 breeds). One breed, the Polish Red, has embryos cryopreserved (more than 1,700) but no semen. The average number of embryos for breeds with embryos (with the exclusion of Polish Red) is 51 (standard deviation 63). For 15 breeds no gametes or embryos are cryopreserved at all.

3.2.5 Management of genetic variation

Maintenance of genetic variation in breeds with small population sizes is a major concern. When the number of breeding animals is low, inbreeding can increase rapidly and genetic variation will be eroded when genetic management programmes are absent. For 66% of the breeds, NCs indicated that there is some kind of inbreeding control, in particular the use of mating planning. More specifically, rotational mating schemes were mentioned for six breeds. For some of the breeds with a population size of more than 1000 females a more classical progeny testing programme (28 breeds) and selection index programme (breeding value or economic index) (22 breeds) was reported. Guidelines and available software for genetic management of small populations are discussed in Chapter 6.

3.2.6 Breed-specific products

Concerning the existence of a specific product directly linked to the local breed, only 27 breeds (25%) mentioned a breed-specific product: 17 for meat, 9 for cheese and 1 for butter. For two breeds, both a specific cheese and a specific meat product was reported.

Marketing and promotion activities have only been properly developed for 17 reported products (out of 27; 63%). The number of breed-specific products appears rather low. However, the survey did not investigate the possible presence of niche food products established with the milk or the meat of the local breed, but that are not breed branded. However, the current low number of breed-specific products reported could imply that development and commercialisation of one specific product directly linked to the breed is not an easy strategy.

3.2.7 Financial support to breeders

Within the framework of the Rural Development Programme, EU countries can decide to financially support farmers rearing farm animals of local cattle breeds with fewer than 7,500 breeding females, that are 'indigenous to the area and in danger of being lost to farming' (EC, 2004). In addition to providing incentives to farmers to reduce the gap in profitability between local breeds and mainstream breeds, EU Member States also have the opportunity to support specific *in situ* or *ex situ* conservation actions (EC, 1999, 2005). Small and Hosking (2010) reported from their survey that a substantial number of European countries use the national allocation from the EU Rural Development Programme to support conservation of Farm Animal Genetic Resources within their jurisdiction. The results of the EURECA survey also indicated that more than 50% of the breeds surveyed are benefiting from direct subsidies to farmers. The subsidy amounts range from €75 to €400/head/year, showing some disparity among the breeds. Our Europe-wide survey also showed that for 26% of the breeds no specific subsidies are given to farmers. As mentioned above, the new EC Regulation provides support for specific conservation actions. This kind of support was not taken into account in the survey, but efficient technical assistance and support may be as important for the maintenance of local breeds as direct financial support to farmers.

3.2.8 Involvement of stakeholders in management of local cattle breeds

Various stakeholders participate in management programmes for local cattle breeds across Europe. We wanted to survey those human and institutional resources (i.e. stakeholders) that can play a major roles in breed conservation. The information can be useful in adjusting or re-directing conservation and development actions. Programmes encountering difficulties could benefit from more successful operations. Table 3.4 reports the involvement of eight stakeholder groups across eight management tasks for conservation of local breeds, in terms of percentage of breeds. Please note that in a specific breed one task can be taken on by more than one stakeholder group.

Breed societies are the most important stakeholders involved in the conservation of local cattle breeds (Table 3.4). Only 14 breeds out of 108 do not have a breed society or a breed interest group. National governments, research institutes and universities also have a strong commitment. A more detailed look at each of the eight management tasks surveyed reveals that:

Table 3.4. Percentage of breeds where the different responsible stakeholders were mentioned for the listed management tasks for conservation of local breeds.

Stakeholders								
	National government	Regional government	Breed society	NGO	Livestock sector organization	Technical institute	University	Others
Breed census	52	10	44	7	1	28	29	4
Breeding programme	27	13	50	3	4	22	19	4
Breed registration	33	29	45	3	8	14	12	2
Ex situ conservation	21	7	27	0	1	15	19	2
In situ conservation	34	14	56	7	0	22	28	4
Performance recording	8	22	35	1	3	0	15	9
Promotion of the breed	35	19	64	22	7	20	26	4
Scientific research	9	7	15	0	1	6	47	2
Technical assistance	19	28	57	7	6	19	38	3

Number of breeds concerned ≥ 25%

Number of breeds concerned ≥ 15% and < 25%

- *Breed census* is mainly a responsibility for the national governments, breed societies, followed by universities, and technical institutes.
- *Breeding programmes* are usually managed by the breed societies. However, other organisations play a role in this task as well, e.g. the largest breeding organisation/AI centre in the Netherlands, CRV, looks after the breeding programme of the MRY breed, CRRG (Centre Régional de Ressources Génétiques) looks after the one for the Bleue du Nord and the Institut de l'Elevage, a technical institute, for the small populations in France. In Serbia, the extension service plays an important role. The conservation programme of the Murdbodner in Austria is managed by a sector organisation and the Magyar Szürke in Hungary by an NGO.
- *Breed registration* is guaranteed by breed societies (45%), national governments (33%) and regional governments (30%).
- *Ex situ conservation* is active for a few breeds. When present (for 33% of the breeds no stakeholders were indicated), breed societies, national governments and universities/ research institutes are the responsible actors.
- *In situ conservation* is a responsibility for breed societies (56%), but also the national government (34%) and universities/research institutes (28%).
- For *performance recording* breed societies are mentioned in the majority of breed cases (35%) as well as regional government (22%). But for 26% of the breeds no stakeholders were mentioned for this task. There is a special case for the Czech breed Ceska Cervinka where performance recording is the responsibility of an NGO and of a sector organisation for the Hungarian breed Magyar Szürke.
- *Breed promotion* is managed by the breed societies in 65% of the cases. However, all other actors participate in this task in 20 to 35% of the breeds, except for the livestock sector organisation (7%).
- *Scientific research* is usually the responsibility of universities/research institutes (47% of breeds).
- *Technical assistance* is provided mainly by breed societies (72%) and then by universities/research institutes (49%).

3.2.9 Positive and negative factors affecting breed development and conservation

NC's were asked to evaluate a list of factors affecting breed development and conservation as very positive, positive, neutral, negative, or very negative. Factors were divided into internal and external (see also Chapter 4 and 7 for more information). Internal factors included breed features, economics of farming and breed management. External factors are related to the attitudes of society, infrastructures and economy. The results of this part of the survey were obtained from 84 breeds.

Polska Czerwona – Polish Red

History

The Polish Red breed is descended from the prehistoric line of short-horned cattle (*Bos taurus brachyceros*). The origins of Polish Red cattle are in the second half of the 19th century, when herds of this breed were established in Polish lands, especially in the south. In the interwar period, the Polish Red breed accounted for 25% of the national cattle population. In the 1960s, there were still about 2 million cattle of this breed, which accounted for 18% of the population. This was followed by a rapid decrease in the population of Polish Red cattle, due to the emergence of more productive breeds and the use of improvement crossing with imported red cattle (mainly Angler). Nowadays, the trend of the population is upwards.

Breeding, conservation and promotion

To protect the breed from extinction, a programme for the genetic resources conservation of Polish Red cattle was started in 1999. The programme is currently coordinated by the National Research Institute of Animal Production. The current population of Polish Red cattle stands at 1,450 animals in 180 herds. By 2013, the number of cows is projected to reach 4,500. The current subsidy per cow involved in the programme is €330.

SWOT

S: Good adaptation to harsh environmental conditions (resistance, undemanding character, good conversion of farm-produced feeds).

W: Lower milk production and lesser quality of beef compared to that of specialised dairy and beef breeds.

O: Presence of EU economic incentives and therefore new breeders joining the Polish Red breed conservation programme.

T: Reduced demand for more expensive organic products due to the crisis.

Internal factors

- Several factors were related to the 'production ability' of the breed. When an evaluation was given about the production capacity, it was usually negative for milk production (40% of the answers) and positive for carcass value (36%). For the functional traits (such as fertility, udder health or feed/legs), the answers were positive in 58% of the cases and very rarely negative. This is consistent with the fact that local cattle breeds are often characterised by their rusticity. It is interesting to note that for the most precise criteria of production (e.g. 'carcass value' and '305d milk production') 25% of the answers were 'do not know/not relevant'. It may be indicative of the lack of knowledge about some local breeds. Furthermore, the factor 'scientific knowledge about characteristics of the breed' was indicated as negative for 38% of the breeds. But the opinions on the quality of the production (milk or meat) were positive in most cases: 74% of the NC's rated quality of meat as positive, and 66% of the NC's rated quality of milk as positive or neutral.
- The 'value of the breed for nature/landscape management (grazing)' was given a very positive or at least positive score in almost all cases (82%), and there was a similar scenario for 'historical /cultural value of the breed' (95%). These two factors illustrate the value of local breeds in valorising their own territory, and highlight their historical and cultural value.
- Concerning more technical factors, 'animal registration and data recording' are mentioned as (very) positive in 75% of the breeds. This high percentage also proves its importance for NCs. Other positive factors, scored by NCs, are: 'current breeding/conservation programme' (67%), and 'presence of breed association and the effectiveness of their activities' (74%).
- The 'farming economic competitiveness' is determined as the main (very) negative factor in half of the cases, which is a worrying development. If the farms that keep local cattle breeds are not seen as competitive, their existence will not be assured.

External factors

- The most 'promising' external factors, according to the answers of the NC's, are 'the availability of infrastructure for management of conservation programmes' (70%), 'the interest of local/regional government in conservation of farm animal genetic diversity' (66%), and 'the availability of public funding for technical assistance' (64%). The 'EU rural policy development' is also seen as a positive factor (58%) as are the 'environmental regulations' (51%). 'Multidisciplinary research on local breeds' is designated as positive for more than half of the breeds (55%).

- The three main (very) negative factors mentioned are not specific for the local breeds but more broadly for agriculture and livestock farming: 'farming sector macro trend' (58%), 'pressure on land use' (48%), and 'availability of local infrastructure for processing (slaughter house, die-cut shop,...)' (41%).

Farmer's quote:

'Local breeds are well adapted to the local environment and do not require high-quality feed'

- Concerning the 'interest of the general public' (i.e. citizens' awareness of the local breed): 40% are valued as positive, 33% as neutral and 23% as negative. But when it was about the 'demand for food quality' or 'tourism related to the breed', the perceptions were much more positive: 58% and 52% respectively, instead of 40%. This underlines the importance of making people aware of the significance of local breeds, particularly through their high quality typical products.

3.3 Conclusions

- A large number of local cattle breeds in Europe are considered to be 'at risk'. Our Europe-wide survey revealed major differences among local breeds in terms of the state of breeds and the state of ongoing conservation actions. This suggests that there is an opportunity to share experiences among breeds and countries.
- Breed conservation can benefit from a large variety of stakeholders or actors, and breed societies, national governments and universities/institutes were reported as the main actors by the NC's that participated in our Europe-wide survey. Strengthened policies and strategies for local cattle breeds will have to take full advantage of these institutions and resources to facilitate the management of local cattle breeds. Besides (other) measures to make breeds more self-sustaining, breeds should take full advantage of EC Regulation opportunities for specific conservation actions.

References

CBD, 1992. Convention on Biological Diversity. www.cbd.org.

Cymbron, T., Freeman, A.R., Isabel Malheiro, M., Vigne, J.D. and Bradley, D.G., 2005. Microsatellite diversity suggests different histories for Mediterranean and Northern European cattle populations. Proceedings of the Royal Society B-Biological Sciences 272: 1837-1843.

EC, 1999. Council Regulation (EC) No 1257/1999 of 17 May 1999 on support for rural development form the European Agricultural Guidance and Guarantee Fund (EAGGF) and amending and repealing certain Regulations.

EC, 2004. Commission Regulation (EC) no 817/2004 of 29 April 2004 laying down detailed rules for the application of Council Regulation (EC) No 1257/1999 on support for rural development from the European Agricultural Guidance and Guarantee Fund (EAGGF).

EC, 2005. Council Regulation (EC) 1698/2005 of 20 September 2005 on support for rural development by the European Agricultural Fund for Rural Development (EAFRD).

European Cattle Diversity Consortium, 2006. Marker-assisted conservation of European cattle breeds: an evaluation. Animal Genetics 37: 475-481.

FAO, 2007. The State of the World's Animal Genetic Resources. FAO, 2007.

Gandini, G.C. and Villa, E., 2003. Analysis of the cultural value of local livestock breeds: a methodology. Journal of Animal Breeding and Genetics 120: 1-11.

Gandini, G.C., Ollivier, L., Danell, B., Distl, O., Georgoudis, A, Groeneveld, E., Martiniuk, E., Van Arendonk, J. and Woolliams, J., 2004. Criteria to assess the degree of endangerment of livestock breeds in Europe. Livestock Production Science 91: 173-182.

Medugorac, I., Medugorac, A., Russ I., Claudia, E., Veit-Kensch, Taberlet, P., Luntz, B., Henry, M. and Förster, M., 2009. Genetic diversity of European cattle breeds highlights the conservation value of traditional unselected breeds with high effective population size. Molecular Ecology 18: 3394-3410.

Negrini, R., Nijman, I.J., Milanesi, E., Moazami-Goudarzi, K., Williams, J.L., Erhardt, G., Dunner, S., Rodellar, C., Valentini, A., Bradley, D.G., Olsaker, I., Kantanen, J., Ajmone-Marsan, P., Lenstra, J.A. and the European Cattle Genetic Diversity Consortium, 2007. Differentiation of European cattle by AFLP fingerprinting. Animal Genetics 38: 60-66.

Small, R. and Hosking, J., 2010. Rural Development Programme Funding for Farm Animal Genetic Resources: A Questionnaire Survey. Report for the National Standing Committee on Animal Genetic Resources, UK.

Toro M.A. and Caballero, A., 2005. Characterisation and conservation of genetic diversity in subdivided populations. Philos. Trans. R. Soc. Ser. B 360: 1367-1378.

Toro, M.A., Fernández, J. and Caballero, A., 2009. Molecular characterization of breeds and its use in conservation. Livestock Science 120: 174-195.

Modenese

History

The Italian Modenese cattle is thought to originate from the heterogeneous cattle population (reddish coat) farmed in the area of Carpi (Modena Province, Northern Italy) in the mid 9th century, with influences from Podolian cattle (grey coat), through some selection for milk (major emphasis), meat (less emphasis), and for a white coat. An official herdbook was started in 1957, and stopped in 1975. In 1986 a new National Register was started, which is still active. The population reached a maximum of 120,000 cows in 1940, and fell to a minimum of 300 cows in 2000. Today the population consists of 650 cows, of which 60% are farmed in herds mixed with Italian Friesian.

Breeding, conservation and promotion

Conservation activities started in the 80's with limited results. These included: inbreeding control, cryo-conservation (semen from 54 bulls is stored for about 20,000 doses), development of a branded Modenese Parmigiano Reggiano cheese made with Modenese milk only and a branded Modenese meat. Two cooperatives for production and valorisation of the branded products were recently created: 'Valorizzazione prodotti bovini di razza Bianca Valpadana Modenese', and 'Bianca Modenese società cooperativa agricola'. Slow Food set up a 'presidium' on the breed. Most farmers benefit from the EU agri-environmental subsidies, approximately €150 per adult cow/year.

SWOT

S: Better functional traits vs. mainstream breed (longevity, fertility, hardiness).

W: Milk and meat production is much lower than for specialised mainstream breed.

O: National and regional interest in local breed conservation and presence of EU economic incentives.

T: National breeder organisation (APA) currently not very interested in the Modenese.

Chapter 4

Viewing differences and similarities across local cattle farming in Europe

Gustavo Gandini, Clara Díaz, Katriina Soini, Taina Lilja and Daniel Martín-Collado

In this chapter:

- Farm characteristics reflecting differences among breeds and countries, for developing local policies.
- Farm characteristics affecting sustainability of local breed farming, common to breeds and countries.
- Profiles of European farmers based on their attitudes towards local cattle production.
- Analysis of strengths, weaknesses, opportunities and threats to be used in developing strategies for breed conservation and development.

4.1 Introduction

European local cattle breeds are distributed across a wide variety of political, social, economic, cultural and environmental contexts. The EURECA project observed across breeds a diversity of both farming structures and methods, and of farmer motivations and values (Gandini *et al.*, 2010). In 1992 the European Union started a policy of economic support for farmers keeping endangered cattle breeds. Under the framework of Council Regulation (EC) No. 1257/99, Commission Regulation (EC) No. 817/2004 provides financial support (incentives) for farmers rearing farm animals of 'local breeds indigenous to the area and in danger of being lost to farming'. The rationale behind incentive payments is to compensate farmers for the lower profitability of the local breeds compared to substituting these breeds with more profitable exotic breeds (e.g. mainstream breeds for dairy and beef). The gap in profitability varies considerably among breeds while incentives are not breed-specific, therefore rearing local breeds remains unprofitable for many farmers (Signorello and Pappalardo, 2003). Opportunities to support conservation measures are to be further strengthened under Commission Regulation (EC) No. 1698/2005. In addition to the possibility for agri-environmental payments to rear farm animals of local endangered breeds, this Regulation also offers new opportunities to Member States to offer specific support for the conservation of genetic resources in agriculture. However, there is a general view that permanent incentives are not a long-term solution for the conservation of animal genetic resources. Therefore, it is necessary to encourage self-sustaining local cattle farming in the near future. For this, three major questions can be posed:

- Are there similarities across European local cattle farming to justify common conservation policies and strategies? Or, do the differences that exist among countries and breeds require policies adapted to specific countries and breeds?
- Should the EU maintain a policy of compensation for the lower profitability of local breeds, and therefore provide general economic support, rather than provide funds for actions capable of accelerating the process toward self-sustaining breeds?
- What should be taken into account, when developing policies to stimulate local breeds towards sustainability?

This chapter summarises some of the work done within the EURECA project, in order to provide answers to these questions, and to provide policy makers and managers with tools for conservation campaigns. In particular, based on data collected in a survey among farmers of fifteen local cattle breeds, we wanted to identify factors affecting the process of developing self-sustaining local breeds, to be used in conservation policies. The analysis aimed to determine factors reflecting both differences and similarities among countries and breeds. A second analysis was carried out to identify farmers' types in order to consider farmers' perceptions when setting up conservation policies. Finally, with the collaboration of breed experts, we identified and examined major internal and external factors that affect local cattle development across Europe.

4.2 Detecting factors affecting sustainability of local breed farming

Until now, information available on European local cattle breeds, including those stored in the EFABIS database (http://efabis.tzv.fal.de), has come from a limited number of people with general knowledge about different breeds. To develop and reorient conservation policies, a wider range of stakeholders should be involved in the process. Farmers keeping local breeds are particularly important actors in conservation because they make the ultimate decision about keeping one or another breed. For this reason it is necessary to understand their values, motivations and expectations for keeping local breeds.

Within the EURECA project we surveyed 371 farmers representing a total of fifteen local cattle breeds in eight European countries. Table 4.1 reports, per country, the names of the fifteen analysed breeds, the breed codes used in the presentation of results, the number of herds surveyed per breed, the population size of the breed and the demographic trends. Most of the breeds were among those classified as endangered following the population thresholds

Table 4.1.Country, breeds and numbers of herds surveyed.

Country	Breed	Breed code	No. of herds analysed	Purpose[1]	No. of cows	Trend[2]
Belgium	Dual-Purpose Belgian Blue	BEBM	23	dual	4,400	s
	Dual-Purpose Red and White	BEPR	18	dual	3,000	d
Estonia	Estonian Native	EEEN	30	dairy	1,500	d
Finland	Eastern Finncattle	FNIS	30	dual	700	i
	Western Finncattle	FNLS	31	dual	3,000	d
France	Ferrandaise	FRFE	19	dual	730	i
	Villard de Lans	FRVI	15	dual	340	s
Ireland	Kerry	IEKE	20	dual	1,200	i
Italy	Modenese	ITMO	26	dual	650	s
	Reggiana	ITRE	30	dairy	1,500	i
The Netherlands	Deep Red	NLDR	21	dual	454	i
	Groningen White Headed	NLGW	22	dual	1,500	s
	Meuse-Rhine-Yssel	NLMR	24	dual	14,400	d,#
Spain	Avileña-Negra Ibérica	ESAN	31	beef	100,000	s,#
	Alistana-Sanabresa	ESAS	31	beef	2,000	i

[1] dual = breeds actively selected for both milk and beef, or unselected.
[2] i =increasing; s = stable; d = decreasing; # = breeds that, although they are not endangered, they experienced severe declines after the fifties.

in Commission Regulation (EC) No. 817/2004 (7,500 cows). We included beef, milk and dual-purpose breeds, and breeds with different demographic trends and structures, in the attempt to capture a good representation of the diversity of local cattle farming in Europe. We worked on a total of fifteen breeds distributed in eight countries, interviewing a rather high number of farmers. However, one has to be aware that there are a limited number of breeds per country (from one to three) and, therefore, when we make inferences about the group of breeds included in each country in the analysis, this may or may not represent the situation of the whole country.

Farmers were asked 42 questions about their farm, farming activities and perceptions of the breeds. In this chapter we analyse 15 aspects that might be related to herd size trends. These aspects are listed in Table 4.2, and are grouped in four profiles: farm profile, farmer profile, economic profile, and social profile. Herd size trend is used here as an indicator of breed

Table 4.2. Fifteen surveyed aspects grouped into four profiles.

Number	Aspect
Farm profile	
1	Farm size (ha)
2	Total number of cows, including all breeds and crosses present on farm
3	% of cows of the local breed present on farm, of total number of cows
4	Herd size relative to the breed average herd size
5	Farmer's evaluation of the local breed, as compared to a mainstream breed on five functional traits
Farmer profile	
6	Age of the farmer
7	Degree of entrepreneurship activity of the farmer, in terms of both activities undertaken in the past and planned for the future
8	Level of cooperation with other farmers of the local breed
Economic profile	
9	% of farm land owned by the farmer
10	% of total family income from cattle farming, including all breeds present on farm
11	% of total family income from local cattle farming
12	% of the cattle production sold on farm and/or on the local market, versus % sold to industry
Social profile	
13	Farmer's opinion on the appreciation of the local breed by society (represented by the following stakeholders: agricultural authorities, research institutes, farmers of main stream breeds, consumers, media)
14	Farmer's opinion on the importance of his breed for society
15	Relevance of tradition for the farmer as the main reason for keeping the local breed

Kerry Cattle

History

It is thought that the Irish Kerry is derived from the little black cow, the Celtic shorthorn, brought by Neolithic man in his migration northwards from the Mediterranean basin. They were first recognised as a breed in 1839 and the herdbook was established in 1887. The number of animals registered per year has fluctuated between 50 and 280 throughout the 20th century. Currently about 300 animals are registered per year. Like many rare breeds, one of the reasons for the decline in numbers has been the displacement of Kerry cattle in favour of other breeds, such as Friesian and Holstein.

Breeding, conservation and promotion

In order to maintain or increase the number of Kerry cattle in Ireland a number of schemes are available to help breeders that keep these animals. A premium of €76 is payable per calf registered in the herdbooks subject to certain restrictions. In addition, under the Rural Environmental Protection Scheme, participants are eligible for a payment of €216 per animal per annum. Semen has also been collected from a number of bulls over the last few years. In 2002, semen from 42 bulls was available (alive or from storage banks). 16% of the available bulls were born before 1985, 15% between 1985 and 1994, 25% between 1995 and 1999 and the remaining 31% were born between 2000 and 2002. The oldest bull with semen was born in 1957.

SWOT

S: Can survive and produce reasonable amounts of high solid content milk in harsh weather and grazing conditions.

W: Small population size.

O: Collaboration with Irish Cattle Breeders Federation for data recording, breeding scheme and conservation strategies.

T: Difficult to acquire animals for semen and embryo collection based on their mean kinship.

farming sustainability. Farmers were asked how the size of their local cattle herd was expected to change in the next five years: increase, no change, or decrease which also includes giving up farming. Thus, we assume that the expected trend in herd size will reflect the number of cows and, consequently, the sustainability of breed farming.

Additional information on the methodology used in the survey can be found elsewhere (Gandini *et al.*, 2010).

4.2.1 Farming local breeds shows remarkable differences among countries

We know that breeds and farming systems can be very different because they are set among a wide variety of historical, political, social, economic, cultural and environmental contexts. These breeds might also differ because they have experienced different influences and pressures in the last few decades, as discussed in Chapter 2.

Then the question, as mentioned in the introduction of this chapter, is: are local cattle breeds in Europe so different that conservation and development policies and strategies have to be adapted to local contexts? And which aspects have to be considered in developing common policies capable that are effective in a general context?

As a first step we analysed differences and similarities among the fifteen aspects surveyed (for details on the statistical analysis used: Gandini *et al.*, 2010). We found twelve aspects, out of the fifteen, that enable the country of origin of farms to be identified, because they describe profiles of farms that are specific to each country. These twelve aspects are listed in Table 4.2, with the exception of: age of the farmer, degree of entrepreneurial activity of the farmer; relevance of tradition as a reason for the farmer to keep the local breed (number 6, 7 and 15). Thus, there is a large variety of situations across cattle breeds analysed in the EURECA project. While average herd size of the two Spanish breeds, Avileña-Negra Ibérica and Alistana-Sanabresa, is respectively 170 and 77 cows, the herd size of the two Finnish breeds, Eastern Finncattle and Western Finncattle, is considerably smaller, respectively 11 and 16 cows. Many farms keep other breeds in addition to the surveyed local breed, both local and mainstream, or crosses. In these mixed herds, the presence of local cattle herds varies from 83% in Western Finncattle, to 28% in the Italian Modenese and 18% in the French Villard de Lans. We also observe a great variation in farmers' motives for keeping the breed.

Further analysis of the twelve aspects that differ among countries reveals that, when combining them, each country has profiles that are mostly country specific. Therefore, the majority of farms from a specific country will not be confused with farms from other countries, except in a few cases. This aspect is graphically represented in Figure 4.1a. For example, looking at the first bar of Figure 4.1a, almost 90% (the green part of the bar) of the farms from Belgium will

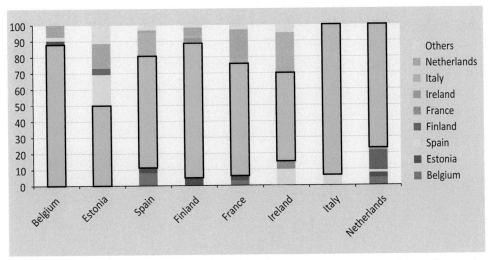

Figure 4.1a. For each country on the horizontal axis, the green part of the bar indicates the proportion of farms from the country that will not be mixed up with farms from other countries, when we consider the twelve aspects that mostly differ among the countries (see text).

never be mixed up with farms of other countries. The remaining 10% show similar profiles to Estonian (the red portion of the bar), Dutch (pink portion), and Spanish (yellow portion) farms.

By completing the image of the farms in the different countries, we obtain a clear picture of how the twelve aspects that are taken into consideration make the farms of a country very similar to the 'average profiles' within the country. Therefore, any common EU policy related to these twelve aspects might have different effects depending on the country of application, and it may benefit farmers from one country and not those of the other countries.

A similar graphical analysis can be performed with the three aspects that were found to be similar among countries (Figure 4.1b) (Finland could not be included in this analysis). These aspects are: age of the farmer; degree of entrepreneurial activity of the farmer; relevance of tradition as the main reason for keeping the local breed (number 6, 7 and 15 in Table 4.2). Attempts to differentiate farms of various countries according to these three aspects will lead to mistakes in most cases. To continue with the Belgium example, the first bar of Figure 4.1b shows that only 5% (orange portion of the bar) of Belgian farms actually have a specific 'Belgian pattern', while the remaining 95% of the farms are mixed up with farms from the other six countries. This is one of the most extreme cases. In general, between 20 and 50% of the farms showed a country pattern that cannot be mixed up with those of other countries, as shown by the orange part of the country bars: 23% of Estonian farms, 47% of the Spanish, 18% of the French, none of the Irish, 21% of the Italian, and 45% of Dutch farms.

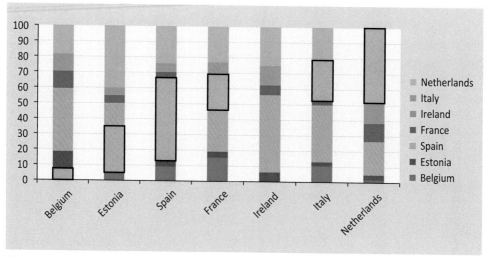

Figure 4.1b. For each country on the horizontal axis, the orange part of the bar indicates the proportion of farms from the country that will not be mixed up with farms from other countries, when we consider the three aspects that are mostly uniform among the countries (see text). The proportion is much lower than when we consider the twelve farm aspects that mostly differ among the countries (comparison with the green part of bars in Figure 4.1a).

The results suggest that policies that are beneficial to local cattle breeds across Europe should try (1) to stimulate degrees of entrepreneurship among farmers, (2) to stimulate the transfer of farms to the next generation, and (3) to value the traditional aspects of local cattle farming.

4.2.2 Factors that affect local breed farming sustainability in all countries

The factors affecting herd dynamics across all European breeds are valuable information, and could be carefully considered in conservation and development policies common to all European countries. Therefore, the next step was to study how common and distinct factors across countries may or may not affect trends in herd size, reflecting somehow the sustainability of breed farming.

We have seen above that most of the aspects studied (twelve out of fifteen) differ consistently among countries. By using the same analytical approach, we found three aspects, among the fifteen listed in Table 4.2, that affect herd size trend in a similar way in all countries. These aspects are: age of the farmer; farmer's opinion on the appreciation of the local breed by society; level of cooperation with other farmers of the local breed (number 6, 8 and 13).

Figures 4.2a-c illustrate relationships between herd size trend (I= increase; N= no change; S= decrease or giving up) and: (a) age of the farmer, (b) farmer's opinion on the appreciation of the local breed by society, and (c) level of cooperation among farmers.

The average age of the interviewed farmers was 48.7 years (SD 11.1; range 25-83). However, the average age of those farmers planning to increase the size of their herd in the next five years is lower (46.3) (Figure 4.2a), and age increases to almost 50 years among farmers

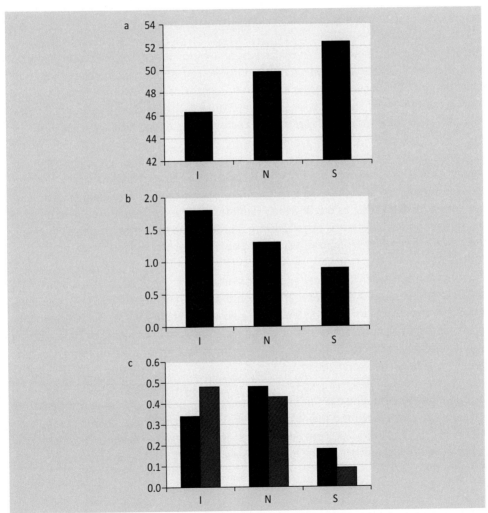

Figure 4.2. Relationship between: the expected change in herd size (I= increase; N= no change; S= decrease or giving up farming) in the next five years and a. age of the farmer; b. farmer's opinion on the appreciation of society for the local breed; c. level of cooperation among farmers (blue = no, brown = yes).

planning no changes in their herd, and to 52.4 among farmers thinking of decreasing the herd size or giving up local cattle farming completely. Therefore, the age of farmer is negatively correlated to our indicator of sustainability of breed farming.

The farmer's opinion on the stakeholders' appreciation of the local breed (measured on a scale from -5 to +5), on the other hand, is positively correlated with herd size trend (Figure 4.2b). Thus, farmers that think society greatly appreciates their breed, plan to increase the size of their herds. As the level of appreciation decreases, we find farmers that plan to maintain the current size of their herds and finally those who plan to reduce their herds or to eliminate the breed from their farms.

The degree of collaboration with other farmers of the local breed also affects the expected herd size trend. Collaboration was measured in two categories: 'collaboration' and 'non-collaboration'. The group favouring collaboration contains mainly those farmers willing to increase their herd size in the near future (Figure 4.2c). On the other hand, it is less representative of those farmers expecting to decrease or abandon local cattle farming.

In general, the farmer's age has been related to delayed or failed generation transfer and higher risk of abandoning farming activities. It could be also related to a less positive attitude towards introducing innovations to their farm. This survey showed how older farmers were more keen on decreasing or quitting their farming activities, or at least not increasing the size of their herd. This effect is observed across breeds, and therefore, strategies favouring the transfer to the next generation will have to be considered in common policies.

The relationship between the farmer's opinion on the appreciation of the local breed by society and his plans for farming activities in the near future is also of interest for EU programmes. We might argue that farmers who feel supported by society are more willing to reinforce or at least continue keeping the local breed. If so, EU programmes aimed at increasing the awareness of society about the positive role played by farmers of local breeds might, in return, benefit local breed development.

We mentioned in the introduction of this chapter that, besides providing compensatory payments to farmers for the lower profitability of the local breed, the EU Commission Regulation (EC) No. 1698/2005 provides support for conservation actions. Considering the importance of collaboration among farmers, specific projects will have to be developed to promote this aspect.

4.3 Identification of the farmer types keeping local cattle breeds

As noted earlier, there are significant differences among farms and farmers keeping local breeds, both within and across countries. A legitimate question is also: could we identify common farmer types regarding the attitudes towards local cattle farming across Europe? To answer this question we analysed the open-ended questions of the farmers surveyed in Belgium, Finland, Italy, the Netherlands and Spain.

Based on a qualitative analysis, three main types of farmers could be identified according to the farming goals: 'Production-oriented', 'Product-oriented', and 'Hobby-oriented'. These three types could be further divided into seven subtypes (Figure 4.3) that carry similarities or differences in the following aspects: attitude towards cattle farming profitability; expertise in farming and need for external support; attitude towards quantity or quality of production; aesthetic values of the animals; commitment to production, and interest in processing and marketing products. The farmers' responses reflected the values of keeping local breed animals.

Production-oriented farmers mainly base cattle farming and its economic profitability on milk and/or meat production. Two subtypes could be found: *sustainable* farmers and *opportunists*. *Sustainable* farmers are typically highly professional in farming and eager to learn about animals and breeding. They consider cattle farming a serious job, and some of them might consider 'hobby farmers' as a threat to local cattle breeding. *Sustainable* farmers are particularly interested in quantity and quality of production. Even though they are aware of and appreciate the quality of products, they are not interested themselves in processing and branding products. *Sustainable* farmers consider that conservation strategies are needed for preserving the good traits of cattle for future breeding, but also to continue the old tradition of their farm. The aesthetic values of the animals arise from good functional traits, like health, longevity, conformation, which make cattle farming sustainable in their view. While

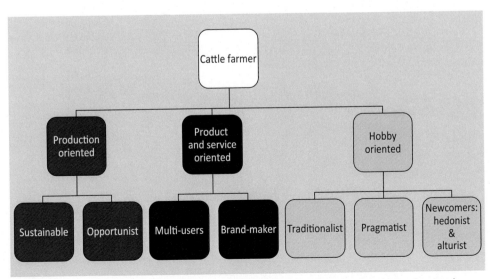

Figure 4.3. Types and subtypes of European cattle farmers keeping local breed animals.

Reggiana

History

The Italian Reggiana cattle was first reported to be farmed in the areas of Parma and Reggio Emilia by monks around the year 1000, and was the most important cattle in these areas in 1809. Pedigree registration started in the 50's and in 1986 the national Herdbook was created. Population size increased to 41,000 cows in 1950, but then the substitution with Brown Alpine and Holstein cattle started. In 1970 the population had dropped to 8,000 cows, and in 1981 the minimum of 450 cows was reached. Since then a positive trend has been observed, also linked to conservation activities, including the production of a specific Parmigiano Reggiano cheese made with Reggiana milk.

Breeding, conservation and promotion

In 1991 a group of passionate breeders created a consortium to add value to the breed by producing the Parmigiano Reggiano cheese. Since its appearance on the market, this branded Parmigiano Reggiano has been welcomed by consumers ready to pay between 30 and almost 100% more for generic Parmigiano Reggiano cheese. Other conservation activities include cryo-preservation (semen from 155 bulls for approximately 17,000 doses) and index selection with control of inbreeding

SWOT

S: High quality of the breed-related product (Parmigiano Reggiano cheese).

W: Small population size: difficulties in breed development, inbreeding risk.

O: National and regional interest in local breed conservation.

T: Risk of having a supply of the breed-related-product that exceeds demand, and a risk of product falsification.

sustainable farmers look for economic profitability in the long term, *opportunists* maximise economic benefits in the short term. Therefore, economic subsidies are extremely important to them, sometimes even the main reason for maintaining or starting to keep the local breed or re-orientating production from one breed to another, even from one species to another. Compared to *sustainable* farmers, *opportunists* are not very keen on cattle farming itself; cattle are a necessity. Cattle enables them to live or to keep the farm in the area. Their main focus is on profitability, whether it comes from quantity, quality, and low production costs or from subsidies. They do not want to put any extra effort into cattle production, and therefore are not interested in processing or branding products, neither do they have any special interest in conservation issues. They may give up farming with a local breed, if economically more attractive ways of earning a living turn up.

Product- and service-oriented farmers acknowledge the unique and multiple values of the local cattle, as a basis for various economic activities, and as a response to new challenges in agriculture. We distinguish two subtypes of these farmers: *brand-makers* and *multi-users*. *Brand-makers* usually have a background in other production sectors, but they are active in gaining expertise in cattle farming and especially in the processing and/or marketing of products. For them cattle farming has to be a profitable business. They put the emphasis on the gastronomic qualities of products instead of production quantity. They actively network with the various actors in the food chain (restaurants, slaughter houses, dairies, shops) and are also active towards the media. They want to promote the preservation of the cattle by processing and marketing products, and they are eager to maintain the niche character of the cattle. In branding, they put the emphasis on the culture and traditions of the region, but always consider the needs and expectations of the modern consumer. Genetic conservation is important as a basis for product quality. Whereas the *brand-makers* are primarily interested in the gastronomic quality of cattle products, the *multi-users* look at local cattle farming more broadly. They see local breeds as elements of 'new' economic activities like tourism, on-farm selling, and nature management. For *multi-users* as well, the quality of the products and services is more important than quantity, but unlike the *brand-makers* they are not interested in processing cattle products on a large scale. Reasons for the conservation of breed diversity cattle range from the genetic to the cultural and ethical.

For the Hobby-oriented farmers the economic profitability of cattle farming is not as important as other factors. There are three subtypes of these farmers: *traditionalists*, *pragmatists* and *newcomers*. *Traditionalists* are typically farmers who have retired from active cattle farming and are very committed to continuing with a local breed. Although they have expertise in farming, the production system might be old-fashioned and they feel they need external support, particularly when it comes to dealing with bureaucracy. Products are used in their household or sold in adjacent neighbourhoods. The main reasons for keeping the breed conservation are the maintenance of the farm tradition and the personal commitment to life-long work with cattle, which they would like to transfer to the next generation. Modern agriculture and lifestyles are often considered a threat to the local breed. *Pragmatists* are professional farmers, whose cattle farming is based on mainstream breeds. However, they recognise the genetic and cultural values of local breeds and they want to contribute to their conservation by keeping a few local breeds among the mainstream breeds, even if it is not economically profitable. Both quality and quantity of the production is not very

Farmer's quote:

'Local breeds are linked to the traditions of the area/the landscape'

important, neither do they have a personal interest in processing or marketing products. At the moment *pragmatics* are not able or not willing to substitute the mainstream breed with the local one, but they might do it in future if mainstream cattle breeding becomes less profitable or if there are any changes in their personal life or at the farm. Finally there are the *newcomers*, who usually have no background in farming. Profitability of farming is not so important. Either these farmers can afford to keep cattle as a hobby or they are willing to keep the cattle even when they have to give up their usual standard of living. For these two reasons, we can distinguish two subcategories under the Newcomers: *hedonists* and *altruists*. For *hedonist* farmers the interest in local cattle mainly arises from personal interests: the cattle fit their lifestyle, cows are just nice and beautiful. These people are not necessarily very committed to keeping the cattle, and the cattle is kept as long as it is easy and nice. *Altruists* think that it is ethically appealing to have local breeds. They want to contribute to 'saving the world' by keeping a local breed. For both *hedonists* and *altruists,* farming practices and bureaucracy, like subsidies and environmental legislation, may cause problems and they need external support. Production or breeding is not so important, but the animals themselves are.

In conclusion, the analysis identified seven farmer types across Europe. Some breeds are clearly connected to one farmer type, whereas other breeds include different farmer types. However, it has to be noted that the types are not country specific, and our analysis showed that many types exist across the investigated countries, even though the number of analysed breed cases per country is relatively low. Obviously specific breeds can favour the existence of certain types of farmers: for example, suckler cows are easier to rear and therefore more attractive for the *opportunists* farmer type. We should also underline that the balance between the types may change, and some of the types may disappear and new ones appear (Soini, 2010, personal communication). More generally, the analysis showed that farmers' reasons for keeping local breeds were rather heterogeneous. This diversity can be considered as a positive factor for breed conservation. If remarkable changes take place (economic, demographic or political, etc.) and one type of farmer gives up keeping local breeds, there are other types of farmers safeguarding the vitality of the local breeds. Therefore, we could say that cultural diversity among the farmers can support the conservation of local cattle breeds, but also vice versa (DPDL/UNEP and DCPI, 2003). Although the diversity of the farmers is important, it provides a challenge to policy makers in terms of finding appropriate solutions in policy, legislation and economic measures for helping the different types of farmers. The diversity of the farmer types makes the system either robust or fragile, depending upon the ability of policies to recognise the heterogeneity.

4.4 Strengths, weaknesses, opportunities and threats affecting cattle diversity

As previously mentioned in this chapter, European local cattle breeds will need to achieve a degree of economic self-sufficiency without external support from public funds. To reach such a state, we need to identify and understand factors that characterise the present state and the future environment where local cattle production will take place. We also need to understand whether or not those factors will show common or breed-specific patterns, because this will shape the development of common, or alternatively breed-specific, conservation policies. Such knowledge will help to establish a decision-making process to identify successful conservation and development strategies. This process is described in more detail in Chapter 7.

4.4.1 Strengths, weaknesses, opportunities and threats

To identify factors that may affect the future of local cattle production across Europe, four open questions were addressed to farmers and other stakeholders of the fifteen breed cases of the EURECA project (Table 4.1), with the exception of Estonian Native and Kerry breeds. People were asked about strengths (S) and weaknesses (W) of the breed, of the farming system and of farmers' characteristics relevant to the development of the local breed. In the analysis, strengths and weaknesses are considered as internal factors of the system (see Chapter 7) that can be controlled by the farmers, who are the main actors in the conservation campaigns. They were also asked about threats (T) and opportunities (O). These refer to the external context, including national and international policies, markets, economic trends, etc. Threats and opportunities can limit or promote the future of the breeds, depending also upon the ability to overcome the weaknesses or to use the strengths when making decisions.

A total of 39 strengths, 28 weaknesses, 18 threats and 23 opportunities were identified from the people interviewed. Internal factors were classified in six categories:
- Animal: productive and functional attributes of animals.
- Breed: aspects related to the population such as size, breed structure, trends, etc.
- Product: quality, uniqueness, etc. of the breed products.
- Farmer: their features, collaboration among farmers, etc.
- Production system: technical, cultural and environmental aspects.
- Marketing system: aspects of the current marketing of breed products under farmers' control, such as branding, distribution channels developed by breeders, etc.

External factors were classified in five categories:

- Market for current products: issues related to the market (mainly about consumer demand) for the current products from the breed.
- Market for new products and functions: Issues related to the market for potential new products or functions, such us landscape management, tourism, etc.
- Production system: aspects related to the competition with high-input high-output production systems.
- Policies and legislation: regulations ranging from subsidies to health, at regional, national or European level.
- Stakeholders: aspects related to the influence of stakeholders.

Table 4.3 lists the categories identified for strengths, weaknesses, opportunities and threats. Concerning the internal environment, products were always, regardless of the country or breed, considered as a strength by both farmers and stakeholders. Therefore, as an example, products could be used either to take advantage of opportunities or to overcome weaknesses in conservation and development strategies. On the other hand, the marketing system of the products was always valued as a weakness for all breeds, independent of the country.

A common pattern across breeds/countries was also observed for opportunities and threats. None of the stakeholders, including the farmers, identified aspects related to the production system as opportunities, but always as threats coming from high-input high-output production systems. Weakness of the marketing system has a different meaning in different countries. For breeds with well-established markets, the risk is related to their capacity to satisfy consumer demands or defend the product from falsification. In other cases, the risk comes from a lack of channels through which products can reach the market. Finally, in general, competition with mainstream products was always envisaged as a threat.

Table 4.3. Categories of strengths, weaknesses, opportunities, and threats identified by farmers and stakeholders.

Strengths	Weaknesses	Opportunities	Threats
Animal	Animal	Market for current products	Market for current products
Breed	Breed	Market for new functions & products	
Products		Policies & legislation	Policies & legislation
Farmers	Farmers	Stakeholders	Stakeholders
Production system	Production system		Production system
	Marketing system		

4.4.2 What are the major strengths, weaknesses, opportunities and threats in each breed?

Table 4.4 presents for each breed case the most important category for *strengths*, *weaknesses*, *opportunities* and *threats*. These priorities were identified by breed experts (see Chapter 7 for more details). All categories identified in the analysis, previously shown in Table 4.3, appear in at least one breed as the highest priority, which again reflects the heterogeneity across local cattle breeds. Strenghts and weaknesses priorities indicate different breed profiles; somehow they also represent different degrees of breed, farming and conservation development. For three breeds out of twelve the major strength is linked to the good features of the animals. In other breeds the major strength is related to the ability of farmers to cooperate and organise themselves into efficient organisations supporting farming activities. Five breeds were envisaged to possess opportunities related to policies and legislation, which basically consist of subsidy availability. On the other hand, for breeds with established markets for their products, opportunities are focused more on the existence of new markets, as for the Reggiana, Modenese and Avileña-Negra Ibérica breeds. For some breeds opportunities are mostly based on markets for new breed functions and products, such as landscape management, as is the case for both French breeds and Groningen White Head in the Netherlands. Perception of the main threat is quite uniform across breeds and countries. For eight breeds their major threat is related to markets for current products from a wide range of perspectives, including competition with other 'local products', risk of product falsification, and for the majority of cases to the generally negative situation of the cattle sector in Europe. In addition to markets for current products, farmers and stakeholders identified both specific policies and lack of support from the administration as major threats.

Prioritising the *strengths*, *weaknesses*, *opportunities* and *threats* in each breed after identifying them results in one important question: what do these differences in priorities mean in terms of developing strategies and policies for conservation? Chapter 7 addresses this question in detail. Based on a general diagnosis of the current situation in the six countries presented above, two common patterns emerge across countries: (1) characteristics of products are commonly perceived as a major strength, while (2) threats mainly come from the market for products from mainstream breeds. As a result, strategies aiming to further develop the marketing of local breed products and functions are key elements in the establishment of common policies. Such strategies may cover (1) development of distribution channels, (2) creation of niche markets for breed-related products and new functions, or (3) increasing social awareness of the roles and functions of local breeds.

Farmer's quote:

'Local cattle breeds can be combined well with other farming activities'

Table 4.4. Main category for strengths, weaknesses, opportunities and threats for each breed. Priorities have been set by breed experts.

Country	Breeds	Strengths	Weaknesses	Opportunities	Threats
Belgium	BEBM	Animal	Animal	Policies and legislation	Policies and legislation
	BEPR	Animal	Breed	Policies and legislation	Production system
Finland	FNIS	Products	Breed	Policies and legislation	Market for current products
	FNLS	Products	Farmer	Policies and legislation	Policies and legislations
France	FRFE	Breed	Marketing system	Market for new function and products	Policies and legislations
	FRVI	Breed	Farmer	Market for new function and products	Stakeholders
Italy	ITMO	Products	Breed	Market for current products	Market for current products
	ITRE	Products	Breed	Market for current products	Market for current products
The Netherlands	NLDR	Farmer	Production system	Stakeholders	Market for current products
	NLGW	Production system	Breed	Market for new function and products	Market for current products
	NLMR	Production system	Marketing system	Stakeholders	Market for current products
Spain	EPAN	Farmer	Animal	Market for current products	Market for current products
	EPAS	Animal	Animal	Policies and legislation	Market for current products

4.5 Conclusions

- Common policies should recognise the existing heterogeneity among European local cattle farming systems, to avoid an imbalance of effects across Europe.
- EU policies, common to all European countries/breeds, should be accompanied by local policies tailored to the specific country/breed cases.
- Common policies capable of helping local breeds to become self-sustaining might include measures to (1) raise social awareness about the positive roles of farmers of local breeds for society; (2) promote collaboration among farmers; and (3) promote the transfer of farms to the next generation.
- Development and conservation policies and strategies should involve a wide range of stakeholders to take into account different views. In particular, farmers should have a key role, because they are among the most important players in the conservation of local cattle breeds.
- Diversity in farmers' values and farming goals should be recognised and taken into consideration when conservation strategies and programmes are designed.
- An analysis of strengths, weaknesses, opportunities and threats for local breed farming is highly advisable to obtain a comprehensive picture when developing conservation actions.

References

DPDL/UNEP and DCPI (eds.), 2003. Cultural diversity and biodiversity for sustainable development. A high-level Roundtable held on 3 September 2002 in Johannesburg during the World Summit on Sustainable Development. UNESCO and UNEP. http://unesdoc.unesco.org/images/0013/001322/132262e.pdf.

Gandini, G., Bay, E., Colinet, F.G., Choroszy, Z., Díaz, C., Duclos, D., Gengler, N., Hoving-Bolink, R.H., Kearney, F., Lilja, T., Mäki-Tanila, A., Martín, D., Musella, M., Pizzi, F., Soini, K., Toro, M., Turri, F., Viinalas, H., the EURECA Consortium and Hiemstra, S.J., 2010. Motives and values in farming local cattle breeds in Europe: a survey on fifteen breed cases. AGRI, in press.

Signorello G. and Pappalardo, G. 2003. Domestic animal biodiversity conservation: a case study of rural development plans in the European Union. Ecological Economics 45: 487-500.

Chapter 5

Role and state of cryopreservation in local cattle breeds

Flavia Pizzi, Delphine Duclos, Henri Woelders and Asko Mäki-Tanila

In this chapter:

- Description of organisational aspects of cryopreservation programmes in Finland, France, Italy and the Netherlands.
- Comparison of sampling strategies adopted in the four countries, and the operation of cryopreservation programmes.
- Identification of internal and external factors affecting cryopreservation programmes, to be used in planning conservation policies.

In situ conservation of local breeds – often regarded as the preferred method for maintaining the variation in local cattle breeds – greatly benefits from being complemented by cryopreservation of genetic material (*ex situ – in vitro*). When *in situ* conservation programmes are not properly planned, breeds may be threatened by inbreeding, by high selection pressure or unbalanced use of some family lineages, by genetic drift or even by extinction. As a backup to overcome such possible shortcomings, cryopreservation of germplasm is not only a very useful tool to maintain genetic diversity within breeds and to support the genetic management of breeds, but also, in the worst case, the cryopreserved germplasm could be used to re-establish the breed.

Guidelines for cryopreservation programmes have been developed in European and global contexts (FAO, 1998; ERFP, 2003; Oldenbroek, 2007). In addition to earlier comparisons between national cryopreservation programmes (e.g. Danchin-Burge and Hiemstra, 2003; Blackburn, 2003), a detailed survey was carried out within the EURECA project to compare cryopreservation activities and policies in Finland, France, Italy and the Netherlands. The purpose of the survey was to detect similarities and differences between these four countries, to compare the countries' strategies with the international Guidelines, and to formulate recommendations for initiating or strengthening cryopreservation programmes.

5.1 Organisational aspects of cryopreservation programmes

5.1.1 History and major developments of cryopreservation

The routine use of artificial insemination (AI) in cattle breeding started after the Second World War in most European countries, whereas the use of cryopreservation for managing local breeds started much later. Finland, France, Italy and the Netherlands recognised the risk of losing breeds and genetic diversity in the 1970s, and followed up with several initiatives. For example, in the Netherlands the Rare Breeds' Foundation was founded in the 1970's and gene banks for some local breeds were established in 1980. A national cryobank started officially in France in 2000. Finland started the National programme for Animal Genetic Resources (AnGR) in 2004, which includes semen storage of Finncattle breeds. And even before that, some less systematic storage work had been done. In the Netherlands, a private foundation developed a national cryopreservation strategy in the 1990s, followed by the government-funded National Programme (CGN) in 2000. In Italy a programme for the characterisation and conservation of the local breeds was developed in 1985; consequently, the breeders' associations started collecting and storing semen of local breeds.

Farmer's quote:

'We got the first cows by accident, but since then we just haven't been able to get rid of them, because they are so cute'

5.1.2 Involvement of stakeholders

A variety of stakeholders is involved in the conservation of local breeds. The involvement and the role of stakeholders in management and conservation of local breeds in Europe are also explained in Chapter 3 'State of local cattle breeds in Europe'. Formally, national governments are responsible for the development of national strategies for the conservation and sustainable use of farm animal genetic diversity. In the surveyed countries, the cryopreservation of animal genetic resources is coordinated by institutes and organisations funded by governments (MTT Agrifood Research in Finland, Institut de l'Elevage in France, Centre for Genetic Resources CGN in the Netherlands, Italian Breeders Association AIA in Italy). These work closely together with a wide range of stakeholders including breed societies, AI organisations, regional governments, farmers' organisations, NGOs such as rare breed societies, research institutes and universities. The survey of the four national cryopreservation programmes showed that AI organisations play an important role in collection (in all four countries) and in long-term storage of semen (in Finland, France, and Italy) of local breeds, in some cases within the framework of the national programme (France, Finland, the Netherlands). In Italy, where a national programme is not yet active, collection and storage of semen falls under the supervision of the local and national breeders' associations. Moreover in all countries, AI centres produce semen of local breeds for routine AI, but also have a commercial interest in the collecting and distributing semen.

5.1.3 Responsibility assessment

The definition of ownership and right of access to the cryopreserved material are important aspects of cryopreservation programmes. In the CGN Gene Bank collection in the Netherlands, all semen is owned by CGN. There are agreements with the organisations or persons and there is a general document describing the conditions for distributing CGN semen. In Finland the decisions about the use of stored semen are jointly made by the national programme and the AI centres. Experts of the Finnish breeding organisation (FABA Service) produce an annual list of bulls recommended for all the registered Eastern and Northern Finncattle cows. In France local breeds' semen is commercialised directly by AI stations except for the doses stored in the genetic reserve. The latter is used only at the request of the French Livestock Institute and of the breed societies. No official agreements exist between the French AI stations and other stakeholders. In Italy the organisations (Italian Breeders' Association or regional governments) that fund semen collection are responsible for the use of semen.

5.1.4 Funding source

Cryopreservation of local breeds in the four countries is mainly funded by national (the Netherlands, Finland, France and Italy) and regional governments (e.g. in France and Italy). Funding for both collection and storage is needed to support cryopreservation. In the

Netherlands, breeding organisations contribute financially to the storage costs; in the case of privately owned collections, the owner or the breeders' organisation pays for storage costs. In Italy collection and storage are financially supported by the Italian Ministry of Agriculture and Forestry and managed by the Italian Breeders' Association. Cryopreservation of Finncattle breeds is funded by the Finnish Ministry of Agriculture and Forestry. Semen collection in France was funded by the French Ministry of Agriculture until the 90s but it is now usually funded by regional governments; the storage of the doses kept by AI stations is funded by the AI organisations themselves.

5.1.5 Use of artificial insemination

Artificial insemination is used for the routine management of reproduction in cattle and is a useful tool for the genetic management of small populations. Nowadays a high percentage of all cows in Finland and the Netherlands, including the local breeds, reproduce via AI. By contrast, in some French and Italian local breeds the proportion of cows serviced by AI is less than 10% (Maraîchine and Casta) or even close to zero (Cinisara, Sarda, Sardo Modicana, Siciliana). According to the results of the Europe-wide survey among the National Coordinators of national programmes on animal genetic resources in Europe (Chapter 3), AI is used less in local breeds than in mainstream breeds: in almost half of the local breeds, less than 20% of cows are artificially inseminated. An increase in the use of AI together with the management of genetic variation could be useful for the conservation of local breeds as explained in Chapter 6 'Assessment and management of genetic variation'.

5.2 Sampling strategies

5.2.1 Aim of storage

Genetic material cryopreserved in gene banks can be used (1) to reconstruct a breed in case of extinction, (2) to create new lines or breeds, (3) as a back-up to quickly modify and/or reorient selection programmes, (4) to support populations conserved *in vivo* in cryo-aided live schemes, or (5) as a genetic resource for research (ERFP, 2003; Oldenbroek, 2007). However, gene banks can also be used for storing alleles of a particular locus that are being eradicated through artificial selection programmes (Fernández *et al.*, 2006). Semen stored in gene banks for the short term can be distributed to farmers to support breeding programmes of local breeds, whereas long-term storage of semen (genetic reserve) is a 'back-up' to secure the breed (Meuwissen, 2007).

The European national gene banks for cattle combine various aims. In Finland particularly, the main aim is to preserve genetic material, semen and embryos, from all the major family

Blanc Bleu Mixte – Dual-Purpose Belgian Blue

History

During the latter part of the 19th century local Belgian dairy cattle were crossed with Shorthorns. At the beginning of the 20th century crossbreeding stopped and until the 1950s selection was strongly focused on milk. Between 1950 and 1970 meat production became more important, and breeders preferred muscular development, discarding milk production. In 1974 the decision was taken to create two separated lines, called the Meat Belgian Blue Breed (BBB) and the Dual-Purpose Belgian Blue Breed (DP-BBB). In 1998, the DP-BBB was considered as an endangered species. Since 2007, the population trend has been stable.

Breeding, conservation and promotion

The muscularity of BBB animals is highly dependent on the presence of the mutation in the myostatin gene (called the *mh allele*) responsible for double muscling. This mutation is admitted in DP-BBB and in 2009 43% of genotyped animals were mh/mh. Calving ease was always given an important weighting in the breeding goal in order to limit the difficult calvings with Caesarean section. Currently, less than 30 bulls are available for AI. All breeders practice AI and 60% of them have bulls for natural mating. The working group also tries to manage the inbreeding.

SWOT

S: The main strength is the comparable profitability with mainstream milk or beef breeds, due to the dual-purpose type and the dynamic and active Breeding Association.

W: The low number of approved bulls for AI and it future impact on inbreeding is considered as a weakness for the breed.

O: One of the main opportunities is the maintenance of the national and European financial support and the regional technical support to DP-BBB breeders.

T: The possible end of the EU milk quota system and/or agri-environmental measures is considered to be one of the main threats for DP-BBB breeders.

lineages within the local breeds. In the Netherlands, the National Gene Bank receives semen from all AI bulls and in addition takes initiatives such as collecting semen from additional bulls. This semen is preferably collected at AI stations, but in exceptional cases it can also be collected 'on farm'. An important aim of the Dutch Gene Bank is to distribute semen for breeding purposes to support the conservation of local breeds 'at risk'. In Italy, where a national cryobank has not yet been set up, semen collection of local breeds is performed by breeders' associations to support the management of local breeds. For each bull a fixed number of doses, at least 50, is kept as a genetic reserve. In France the majority of the semen of local breeds is collected by AI centres for routine use, but approximately 200 doses per bull are kept for management of conservation programmes. Some doses are also stored in the French National Cryobank as genetic reserve.

5.2.2 Type of genetic material stored

Semen and embryos form the most common material for the cryopreservation of farm animal genetic resources. Both types of germplasm have possibilities and limitations. Embryos are valuable material for cryopreservation because they carry the entire genome, including the extra-nuclear genetic material contained in mitochondria. The storage of embryos would allow the complete and immediate recovery of the breed in case of extinction (within one generation). In contrast, for reconstruction of the breed by using semen alone, several generations of repeated backcrossing are needed, and consequently the required number of semen doses is high. For this reason and in particular for species with a long generation interval, such as cattle, the storage of both semen and embryos is recommended (Boettcher *et al.*, 2005). In addition, the costs for storing embryos plus semen of cattle in gene banks are not significantly different from those for storing only semen (Gandini *et al.*, 2007).

The most commonly cryopreserved genetic material of European local breeds is semen, as shown in the four countries studied, as well as in the Europe-wide survey carried out within the EURECA project (described in Chapter 3). The latter shows that in 93 out of 108 breeds only semen is stored, whereas only in 26 out of 108 breeds are both semen and embryos stored. In the four countries studied, embryos are also stored for the Finncattle, the Dutch local breeds – Dutch Friesian Red and White and Deep Red – and for some Italian local breeds. For almost all the bulls DNA samples are also available, and stored as genetic reserve.

5.2.3 Selection of donors

The selection of donor animals for cryopreservation can be based on different criteria: (1) random sampling, (2) selecting animals carrying specific genotypes/alleles/haplotypes, or (3) maximising genetic variation (e.g. Oldenbroek, 2007). Sampling strategies vary in the four countries surveyed. Most of the time, the choice of bulls of which semen will be collected is based on pedigree information, morphology and performance (of relatives). Finland has made investigations about the optimum policy for the management of storage: the impact of each bull on the genetic variation is considered with the aim of minimising the average coancestry among the stored material. In the Netherlands, the sampling strategy for cattle breeds largely relies on the activities of AI organisations (sample of 100 straws of each AI bull goes to gene bank), or on proposals of breed interest groups or herdbooks. However, the National Gene Bank (CGN) will invest in depositing additional semen only after considering coancestry. In Italy bull donors are chosen based on their contribution to increasing genetic variation of the stored genetic material. Donors are mostly selected by the Italian Breeders' Association, or by the Istituto di Biologia e Biotecnologia Agraria of CNR in the specific case of the Lombardia AnGR cryobank. In France, bulls to be collected are chosen by the French Livestock Institute in collaboration with breed societies. The most important criterion is to have a good representation of the existing genetic variability, while bull morphology and characteristics of the bull's dam are also considered. Decisions are based on pedigree analysis.

5.2.4 Number of sires and number of doses in storage

For a total of 52 local breeds from Finland, the Netherlands, France and Italy detailed information on 2,536 bulls was collected. Table 5.1 reports the status of endangerment of the 52 surveyed breeds, following FAO criteria (i.e. populations with fewer than 1000 breeding females are considered to be endangered, and those with <100 breeding females in the population are regarded as critical).

We observe two sampling strategies in the four countries: (1) few bulls with many doses are stored over a medium time period (e.g. in Finland and France), and (2) semen of as many as bulls as possible is collected but with fewer straws and rapid turnover (e.g. in Italy and the Netherlands). As shown in Table 5.2, in the Netherlands and Italy, the majority of bulls of

Table 5.1. Local cattle breeds cryopreservation data.

Country	Breed classification according to FAO criteria					
	Critical	Endangered	Critical maintained	Endangered maintained	Not at risk	Total
Finland	0	2	0	0	1	3
France	1	0	1	12	4	18
Italy	0	1	1	9	12	23
Netherlands	1	3	0	0	4	8
Total	2	9	4	17	19	52

Table 5.2. Distribution of number of semen doses per bulls.

Number of doses/bull	< 200 doses	200-1000 doses	>1000 doses
Finland	19%	36%	45%
France	24%	17%	59%
Italy	57%	40%	3%
Netherlands	64%	28%	8%

local breeds have less than 200 doses stored as genetic reserve. Conversely, in France and Finland, almost 50% of the bulls have more than 1000 doses collected. In France when a bull is taken to the AI centre for semen collection the aim is to produce 3,000 doses.

5.2.5 Birth date of bulls

In most countries, almost a quarter of the bulls stored were born before 1980, except in Italy, where there are only a few animals in this age category (Table 5.3). The distribution of year of birth of the bulls in each country reflects the evolution of the cryopreservation programmes. In those countries where the national programme started in the early 1980's (e.g. Finland, France, and the Netherlands), we observe a high percentage of bulls born before 1999, while in Italy, where cryopreservation of local breeds started later, most of the bulls represented in the storage were born in recent years. In France few bulls born after 2000 have been stored because there are already many doses per breed and storing new bulls is no longer a priority for some breeds.

Table 5.3. Year of birth of bulls stored in national gene banks.

	<1980	1980-1999	>2000	Unknown
Finland	22%	50%	28%	
France	26%	61%	17%	
Italy	3%	58%	34%	5%
Netherlands	26%	55%	19%	

5.3 Operations of cryopreservation programmes

5.3.1 Collection methods

Biosecurity is an important issue for gene banks because gametes and embryos may carry pathogens capable of surviving cryopreservation. Because a risk assessment of transmission through gene banks is often not performed, the recommendations of the Terrestrial Animal Health Code of the World Organization for Animal Health (OIE) should be applied. Nevertheless, meeting the requirements for local breeds is sometimes difficult, causing problems for the establishment of cryobanks.

Semen can be collected in EU-certified AI centres complying with all the pertinent regulations that exist at the time of collection, or it can be collected *on farm* if derogation with regard to the procedures exists within the country. The sanitary status of semen may differ depending on the semen collection procedures.

Semen collected in the EU-certified AI centres has the same status as semen of mainstream bulls and can be stored without any special conditions. If necessary it can even be exported to other countries. However, the costs of doses produced at AI centres can be very high for a local breed when only a small number of doses per bulls is collected.

Collection *on farm* is less expensive but the semen has a lower sanitary status compared to semen collected in an AI centre. In this context, special attention must be paid to the sanitary condition of the bull, the farm and the place where semen is collected.

In Finland and France, all semen is collected in EU-certified AI centres. In the Netherlands, the majority of bulls are collected in EU-certified AI centres but CGN has a special derogation to collect semen *on farm* that may be used, under strict conditions, to support the breeding programme for local breeds with fewer than 7,500 females. In Italy, 66% of bulls of local breeds have been collected *on farm* by taking advantage of special derogations. If we consider the Italian breeds with fewer than 1000 females, almost all semen has been collected on farm. The collection methods used in the different countries/breeds are summarised in Figure 5.1.

In small populations where few doses per bull are needed for conservation (see Chapter 6. 'Assessment and management of genetic variation') the storage of semen collected *on farm* under strict control could be an option.

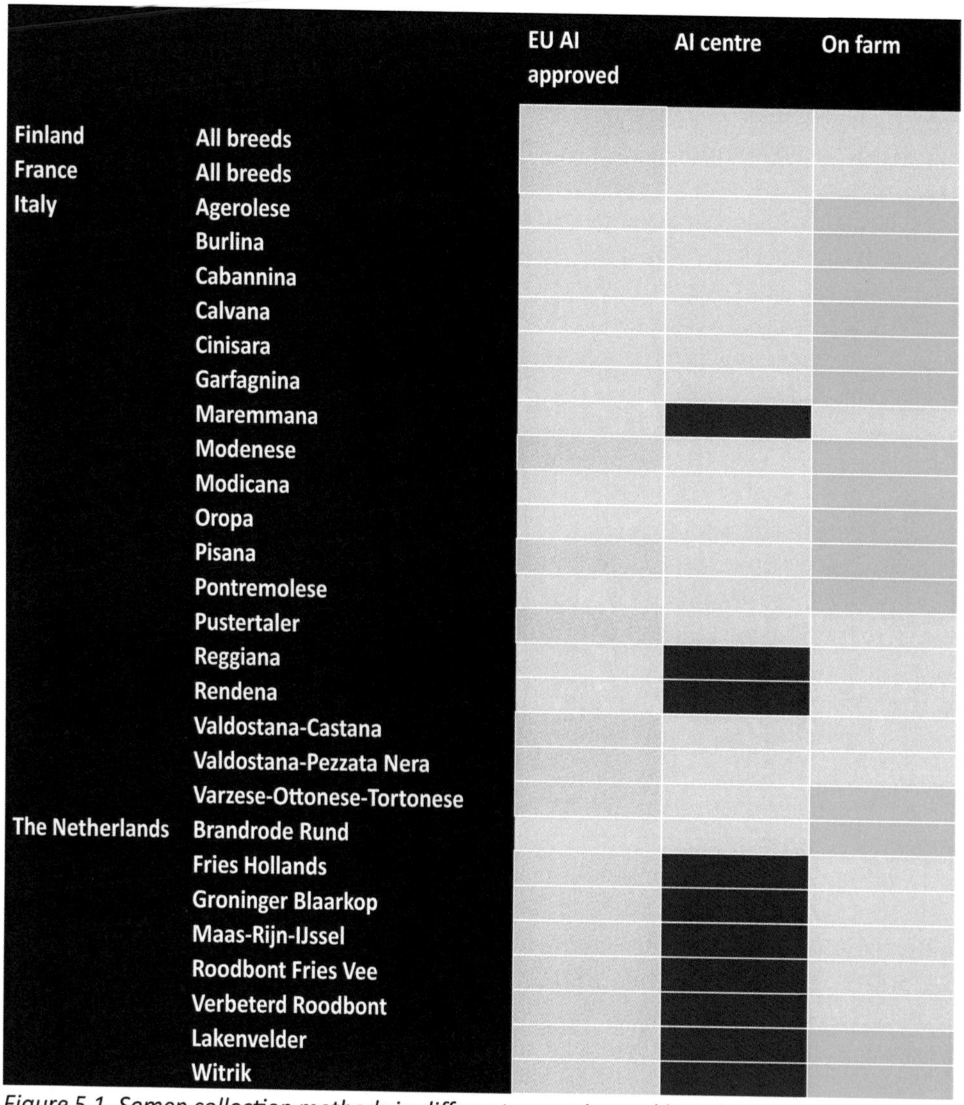

		EU AI approved	AI centre	On farm
Finland	All breeds			
France	All breeds			
Italy	Agerolese			
	Burlina			
	Cabannina			
	Calvana			
	Cinisara			
	Garfagnina			
	Maremmana			
	Modenese			
	Modicana			
	Oropa			
	Pisana			
	Pontremolese			
	Pustertaler			
	Reggiana			
	Rendena			
	Valdostana-Castana			
	Valdostana-Pezzata Nera			
	Varzese-Ottonese-Tortonese			
The Netherlands	Brandrode Rund			
	Fries Hollands			
	Groninger Blaarkop			
	Maas-Rijn-IJssel			
	Roodbont Fries Vee			
	Verbeterd Roodbont			
	Lakenvelder			
	Witrik			

Figure 5.1. Semen collection methods in different countries and breeds

5.3.2 Semen packaging

A straw with 0.25ml is the most common package used for semen from local breeds, but in Italy a high percentage is packaged in 0.5ml straws. In the past, 1ml straws and pellets were used. The latter were abandoned because they do not fulfil the requirements of sanitary security and clear identification, fundamental for the long-term storage.

5.3.3 Storage sites

As stated in the international Guidelines (FAO, 1998; ERFP, 2003), genetic material from a single bull must be stored in two different well-secured sites for security reasons. This condition is essential in the case of local breeds, where the material has a high conservation value. Most of the semen from European local breeds is stored in two sites fulfilling this requirement. If only one site is available, then the semen must be split into at least two tanks to avoid the risk of losing the material due to accidents, fire, etc. The sites can be located at the headquarters of the cryobank (the Netherlands and France), or at an AI centre (Italy and Finland), or at University and Research Institutes (the Netherlands and Italy). In the Netherlands, CGN has one storage site for the EU certified semen and two for the non-certified semen (e.g. collected *on farm*).

5.3.4 Pedigree information

In most cases (86%) the pedigree of the bulls collected in cryopreservation programmes is known. In Finland, only the oldest and more exotic family lineages have animals with unknown parents. The pedigree can also be retrieved from the database of the breeding organisations.

5.3.5 Cryobank data management

Cryobanks require comprehensive documentation of their content. Different systems can be used for the management of cryobank data, including CryoWEB, a web application for national genebank management developed by the Institute of Farm Animal Genetics (FLI), Mariensee, Germany. In the EFABISnet project supported by EU, CryoWEB has been installed in 10 countries, including Finland, Italy and the Netherlands, to serve as the information centre of national gene banks. In Finland, the data from the cryo-material for cattle breeds is maintained in the FABA Service database. In France, all genetic material stored in the French National Cryobank is registered in a specific database called 'Cryobase'.

5.4 Factors affecting cryopreservation programmes

The experts involved in the cryopreservation programmes of Finland, France, Italy and the Netherlands were asked to identify and rank the key factors affecting the state of AnGR cryopreservation in their country. The aim was to have a clear picture about the current situation of cryopreservation for the development of national programmes and to identify and design efficient policies at national or European levels.

The identification of the key factors has been approached in two stages, following the methodology in Chapter 7 'Using decision making tools for the development of breed strategies'. First, experts defined the internal and external factors affecting the cryopreservation of AnGR within their country. Internal factors are the strengths and weaknesses of their programmes that have affected the past and may compromise or benefit the future of programmes. External factors are the threats and opportunities of the national framework that could limit or favour the efficient development of AnGR cryopreservation. Second, experts were asked to rank the factors from 1 (least important) to 6 (most important) to quantify their importance. The experts agreed on some common internal and external factors that affect the various levels of the cryopreservation programmes in the four countries (Table 5.4).

Table 5.4. Common factors affecting cryopreservation programmes.

Strengths and weaknesses	Opportunities and threats
Funding	Farm AnGR conservation policies
National coordination	Support by breeders' associations
Breeders' association involvement	Infrastructure: AI centre (number, geographical
AI centres' involvement	distribution, etc)
Regulatory framework	Funds for AnGR management
Genetic material stored	Socio-cultural interest in AnGR conservation

5.4.1 Strengths

The experts of Finland, France and the Netherlands considered the initial financial support from central government (Ministries of Agriculture) and the presence of a national programme as the principal strengths for the development of efficient cryopreservation programmes (Figure 5.2). Active breeders' associations are considered a driving factor in the Netherlands, France and Italy. The technical involvement of AI centres in the cryopreservation of local breeds is recognised as a positive factor for the French, Finnish and Italian programmes. The presence of stocks of semen previously collected and their use in supporting local breeds' conservation is perceived as a highly positive internal factor, mainly in Italy. Concerning the current framework, the use of EU common regulations is considered as a strength for Finland and France. Italian and French experts underlined the possibility of some derogations for local cattle to facilitate collection (e.g. collection *on farm*, complete pedigree not compulsory) as a major strength.

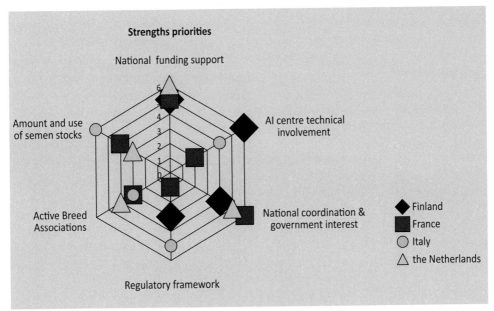

Figure 5.2. Ranking of cryopreservation strengths as indicated by national experts. Rank 1 is least important, rank 6 is most important.

5.4.2 Weaknesses

Limited availability of funds is considered as a weakness of cryopreservation programmes. In Finland in particular, the high cost of cryopreservation was mentioned and Italian experts reported limited public financial support. The absence of commercial interest (value) in semen of local breeds was considered (indirectly) to be a limiting factor in the Netherlands (Figure 5.3). In all four countries there is concern about the amount and type of genetic material stored in the cryobanks (e.g. limited number of local breeds have embryos stored). Lack of regulation for the protection of doses collected is perceived by the French experts as a potentially relevant weakness, and, at a lower level, the ambiguous ownership of the genetic material is perceived as a weakness by the Finnish experts. The Dutch experts recognise the lack of structured breeding programmes for local breeds as a weakness for the cryopreservation of AnGR in their country. According to the Italian experts, the cryopreservation activities greatly suffer from the absence of a national coordinated programme.

Groninger Blaarkop – Groningen White Headed

History

The Groningen White Headed is a native Dutch breed. The first descriptions of this breed go back to the 14th century. Already in the Middle Ages portraits of red and black White Headed cows were being painted. Friesian cattle traders bought White Headed cattle for slaughtering on a cattle market in London at the end of the 19th century. At the start of the 20th century, 90% of all cattle in the Province of Groningen consisted of White Headed cattle. White Headed cattle were also bred in the Province of Zuid-Holland, around Leiden, and along the Rhine in Utrecht. The number of pure-bred cows is less than 1000 but the population trend has been increasing slightly over the past few years.

Breeding, conservation and promotion

There are a number of national and regional breed interest groups. One major breed interest group, called the 'Blaarkop Stichting', is very active in promoting the breed. In the mid '70s most farmers started using Holstein Friesian (HF) semen from the United States. By crossbreeding White Headed with HF, the milk production increased but the typical appearance of White Headed lasted for some generations. The recent increasing interest in functional traits, like feed efficiency, fertility and health, may offer new opportunities for the Groningen White Headed. Semen from 49 Groningen White Headed bulls, collected between 1973 and 2005, is conserved in the National Gene Bank (CGN).

SWOT

S: The strengths for the White Headed cattle are their good performance on functional traits and also the productivity and popularity of in particular the (F1-)crossbreeds.

W: There are a limited number of good AI-bulls available, and the genetic variation within the breed is a point of concern.

O: The biggest opportunity for the Groningen White Headed cattle is the new interest in functional traits and therefore in the use of White Headed sires for crossbreeding with Holstein Friesian.

T: The expected abolishment of the milk quota, combined with environmental legislation and further emphasis on increasing the efficiency of milk production are considered as threats.

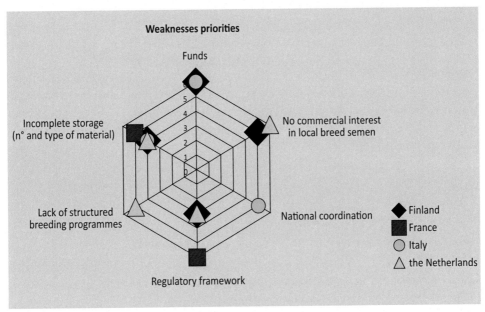

Figure 5.3. Ranking of cryopreservation weaknesses as indicated by national experts. Rank 1 is least important, and rank 6 is most important.

5.4.3 Opportunities

The use of *ex situ* (cryopreservation) strategies to support the *in situ* conservation programmes represents the main opportunity for an efficient development of cryopreservation programmes, as shown by the high ranking given to the presence of AnGR conservation policies and to breeding programmes for local breeds by all the experts (Figure 5.4). External factors like the presence of local breeders' societies in all four surveyed countries, and an increased awareness of the conservation of local breeds, mainly in Finland, are recognised to have a positive effect on cryopreservation. The involvement of breeding companies and AI centres in taking joint responsibility for the long-term storage of local breeds is perceived as an opportunity to facilitate cryopreservation activities in France, Italy and the Netherlands.

5.4.4 Threats

Negative trends in public funds, together with the high costs of collection and storage of genetic material (Finland), are recognised as the major threats for the development of cryopreservation programmes (Figure 5.5). The lack of national strategies, or their discontinuity, are also considered as a limiting factor by the Italian and French experts. The uncertainty about support from AI centres in terms of their continued interest in local breeds (the Netherlands) and about the availabilty of AI expertise in the region where local breeds are raised (Italy),

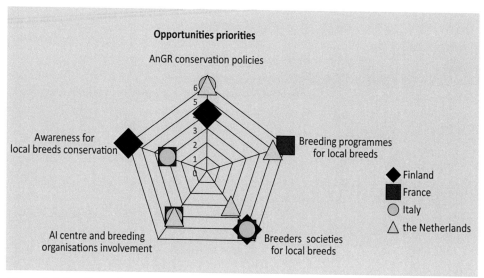

Figure 5.4. Ranking of cryopreservation opportunities as indicated by national experts. Rank 1 is least important, and rank 6 is most important.

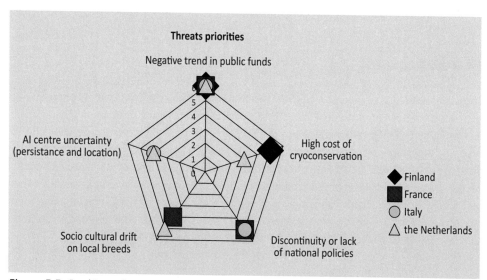

Figure 5.5. Ranking of cryopreservation threats as indicated by national experts. Rank 1 is least important, and rank 6 is most important.

can threaten the cryopreservation of local breeds. Socio-cultural and demographic factors, in terms of a decreasing interest in local cattle farming (e.g. high farmer age in the Netherlands), are generally perceived as a threat to cryopreservation programmes.

5.5 Conclusions

- Cryopreservation strategies can greatly benefit from being developed within the framework of a national programme for AnGR conservation. There is always a need for national or public or joint funding for national cryopreservation programmes with long-term objectives.
- Countries have different ways of organising cryopreservation programmes at the national level, depending on the role and responsibilities of different stakeholders. However, in all four countries studied, the close involvement of breeders of local breeds, breed associations and AI centres in linking the cryopreservation schemes with routine AI operations, was considered as the most important factor for the development of efficient cryopreservation programmes. The collaboration between stakeholders should be confirmed by contracts.
- The inventory in these four European countries gives an indication about substantial semen gene bank collections, developed from before 1980 until today. Freezing semen is a cost-efficient method to preserve variation; choosing bulls for storage should be preceded by careful planning and optimisation of genetic contributions. Embryo collections are less common, while embryo storage in combination with semen could be a (cost-)effective strategy for re-establishing the pure genomes of a breed.
- Cryobanks require comprehensive documentation of their content, donors and freezing procedures. A database system for the management of all this information should be implemented within a cryopreservation programme.
- Long-term storage and the use of cryopreserved material of local breeds requires specific security and sanitary measures. Secured storage of genetic material should be guaranteed through duplicate storage sites and continuous monitoring of the storage conditions (e.g. sufficient liquid nitrogen in the tanks). Countries should develop and implement a national regulatory framework to control the sanitary risks but at the same time facilitate flexible collection (in particular *on farm*) and the use of genetic material of local cattle breeds.

References

Blackburn, H.D., 2003. Conservation of U.S. Genetic Resources through Cryopreservation. In: Planchenault, D. (ed.) Workshop on Cryopreservation of Animal Genetic Resources in Europe, Paris, France.

Boettcher, P.J., Stella A., Pizzi, F. and Gandini, G. 2005. The Combined Use Of Embryos And Semen For Cryogenic Conservation Of Farm Mammal Genetic Resources. Genetics, Selection, Evolution 37: 657-675.

Danchin-Burge, C. and Hiemstra, S.J., 2003.Cryopreservation of domestic animal species in France and the Netherlands – Experiences similarities and differences. In: Planchenault, D. (ed.) Workshop on Cryopreservation of Animal Genetic Resources in Europe, Paris, France.

ERFP, 2003. Guidelines for the Constitution of National Cryopreservation Programmes for Farm Animals. Hiemstra, S.J. (ed.) Publication No. 1 of the European Regional Focal Point on Animal Genetic Resources.

FAO, 1998. Management of small populations at risk. Secondary guidelines for development of national farm animal genetic resources management plans, FAO, Rome, Italy.

Fernández, J., Roughsedge, T., Woolliams, J.A. and Villanueva, B., 2006. Optimization of the sampling strategy for establishing a gene bank: storing PrP alleles following a scrapie eradication plan as a case study. Animal Science 82: 813-821.

Gandini, G., Pizzi, F., Stella, A. and Boettcher, P.J., 2007. The costs of breed reconstruction from cryopreserved material in mammalian livestock species Genetic, Selection, Evolution 39: 465-479.

Meuwissen, T.H.E., 2007. Operation of conservation schemes. In: Oldenbroek, K. (ed.) Utilisation and conservation of farm animal genetic resources. Wageningen Academic Publishers, pp. 167-194.

Oldenbroek, J.K. (ed.), 2007. Utilisation and conservation of farm animal genetic resources. Wageningen Academic Publishers, 232 p.

Villard de Lans

History

The French Villard de Lans is a cattle breed that originates from the Vercors Mountains not far from the city of Grenoble, in the French Préalps. The Villard de Lans breed was maintained in its area of origin until the end of the 50's. The decline in bovine labour was fatal for the breed, as was the aggressive disease prevention against tuberculosis and brucellosis and the installation of a policy of specialisation of the farms. At the beginning of the 20th century, the number of females totalled around 16,000 cows, falling to just over 100 cows at the beginning of the 80's.

Breeding, conservation and promotion

Today it is not possible to associate Villard de Lans breed with a specific product or breeding system. A little more than half of the animals are bred in suckling herds, sometimes by part-time breeders and, in general, products are sold on local markets. The other animals are bred in dairy herds. Some herds are used for a branded cheese product: AOC 'Bleu du Vercors-Sassenage'. Bulls of different origins have been found and collected since 1977. Today, 27 bulls are available for AI with a good genetic diversity which prevents inbreeding (total inbreeding of females is 3.9%) and in 2008 there was a total of more than 400 females.

SWOT

S: The historical link between breed and territory is strong and there are some passionate breeders.

W: Production of the breed is lower than the mainstream breed, and cooperation among farmers is limited.

O: There is a regional and national interest in the breed, and it should be possible to develop niche products.

T: There is competition from tourism and urbanisation in this territory.

Chapter 6

Assessment and management of genetic variation

Asko Mäki-Tanila, Jesus Fernandez, Miguel Toro and Theo Meuwissen

In this chapter:

- Description of different kinds of estimates of genetic variation that are available.
- Discussion about the differences between molecular and pedigree methods.
- Description of how to compare populations for state of variation.
- Discussion on what is the best parameter for management of variation.
- Discussion on what we can gain from the mating strategies.
- Conclusions on how we can support the management schemes with cryopreservation.

6.1 Cattle breed diversity

The previous chapters have described how to make local breeds more self-sustaining through development and conservation actions. Most local breeds are the result of particular adaptations to singular, sometimes hard, environments. In many cases, no other breed could survive in that habitat. The extinction of the local breed would break the equilibrium reached by the coexistence over very long time periods and would increase the dependency on external products. European rural areas and cultures are closely linked to local livestock breeds, which play a key role in economics by providing employment and trade, including export markets. Cattle also make a recognised contribution to landscape management.

In many cases, cattle breeds differ from a general pool of breeds with one or two special traits. There could be cattle breeds with a high frequency of the kappa casein β allele which has desirable cheese-making properties. Some beef breeds may be famous for the marbling properties of the meat. But before the unique features are recognised, we need a thorough characterisation of the breeds.

The maintenance of diversity could be considered at the species level, where the number of different breeds is a measure of the diversity. Within species, breeds are like homogenous lines and the metapopulation formed by the breeds carries lots of genetic variation with a low risk for the erosion of the total diversity. The metapopulation concept and its properties are understood when we realise that breeds are formed via redistribution of original variation in the ancestral base population. Therefore, the within-species diversity is another worthwhile reason for keeping local breeds.

From a production point of view, the increasing demand for products from cattle, like milk and meat, is mainly covered by the mainstream breeds. Those breeds show higher efficiency and better performance under standard production systems, as stated in previous chapters. However, the high selection pressure imposed on the mainstream breeds may rapidly, or at least gradually, erode their genetic diversity. Consequently, some problems are arising from performance in fitness-related traits, often antagonistically related to production traits and sensitive to inbreeding depression. Moreover, the lack of genetic variability may compromise the long-term progress in selected traits and prevent the adaptation of the breeds to new market scenarios or changing environments (e.g. climate change).

In this context, local breeds may be considered as a source of new material to assure the viability of mainstream production system, compensating for the loss of alleles caused by selection. Besides this, the modern European citizen regards food security and quality as very important criteria in the development schemes of animal production. This also makes room for the use of local breeds with exceptional milk or meat characteristics. Therefore, the maintenance of small autochthonous breeds appears to be a key element in guaranteeing the future of the overall system.

Local breeds are usually maintained with low census sizes, which puts them at risk of disappearing. Consequently, the genetic diversity stored in each of them should be treated with great care, especially if we believe that they bear unique genes that we do not want to get mixed with that of other breeds.

Farmer's quote:

'Local cattle are extremely suitable for grazing management in nature reserves. Good use of natural pastures all year around'

6.2 Importance of genetic variation

Levels of genetic diversity are connected with the adaptability of the population and, thus, with the ability of the breed to respond to selection. The response to artificial selection of a trait is directly proportional to the amount of genetic variation in the trait. Furthermore, the response to natural selection pressure (i.e. changing environments) is equal to the genetic variability (Fisher, 1958).

Genetic variation is also related to the fitness of individuals: (1) in the short term, smaller populations develop more inbreeding and, consequently, suffer more from inbreeding depression in both fitness and production traits; (2) in the long term, the accumulation of deleterious mutations is higher in populations with low census numbers and, thus, low levels of diversity. All the statements above clearly indicate the risk of extinction a breed will be exposed to if we do not manage its genetic variation carefully.

When dealing with the conservation of a local breed we have to consider a two-step approach. First we have to assess the present state of the population. This measure of diversity will allow us to determine the degree of endangerment of the breed and will also provide some information on the history of the population. As such, we can find out what was wrong with its past management and try to correct it. The design of the programme is very similar whether we have a conservation or selection programme. In worst cases, the analysis may reveal that the actual basis of genetic variation in the population is very narrow, either due to a small census size, recent bottleneck or highly unbalanced use of elite individuals in intensive selection. Immediate action should be taken to stop any undesirable development and to initiate operations to prevent further decline in variation. If uneven genetic contributions are not corrected within a few generations, they cause a permanent bottleneck.

Next we have to decide on a management strategy to keep the population in a 'healthy' state. Overall, the management of genetic variation would need planning for appropriate design and operational tools. Without planned management in a genetically narrow population there will be immediate consequences such as increased homozygosity, and, in the long-term, the poor management results in reduced potential for genetic change and the danger of an accumulation of deleterious mutations. For example, in the case of a new infectious disease, we may envisage the risk that a population with less variation may be severely affected if it lacks resistant variants.

The genetic impoverishment of populations does not apply exclusively to local/small breeds. In many cases, the high selection intensity in mainstream breeds (e.g. dairy cattle) has been narrowing the genetic variability of the population. The active international exchange of semen coupled with efficient marketing has led to a situation where the current bulls used for artificial insemination (AI) are descendants of very few ancestral sires (e.g. Wickham and Banos, 1998). As a consequence, there are only a few dominant families and their genetic profile determines the genetic variation. Consequently, management of these commercial populations should also account for the maintenance of the genetic diversity.

Although in this chapter we will focus on within-breed variation, between-breed variation is equally important. While, traditionally, a close relationship between breeds was deduced from the similarities in the exterior traits (shape, colours, shape of horns, etc), molecular genetics now provides cheap and reliable tools for a more precise assessment of the genetic differences. Even assuming that, in principle, we are not intending to merge breeds, these parameters are useful in ranking breeds according to their deviation from the general pool of the breeds, at the same time considering the variation present within the breed (Boettcher *et al.*, 2010).

The general aim of genetic conservation is to maintain within- and between-breed diversity. Within-breed diversity is important for the genetic adaptation of a population to changes in the production and economic environment, and for preventing inbreeding problems. Between-breed diversity is important for providing alternative gene pools if a breed happens to run into genetic problems due to genetic drift or if, due to changes in the production system, traits are required which a breed lacks.

Genetic measures of differentiation (distance) between breeds can be used to take decisions about the common management of them, for example for 'rescuing' one of the breeds by receiving animals from another breed (ideally a breed that is closely related). Those parameters can also be used to rank breeds in case we have to reduce the number of breeds we can keep, or if we want to create a synthetic breed by mixing several breeds (Toro *et al.*, 2009).

6.3 Describing the state of variation

6.3.1 Direct molecular methods

Genetic variation in a population can be expressed in different ways. Although, usually, we are interested in morphological or performance traits of animals, we must realise that differences

Itäsuomenkarja, kyyttö – Eastern Finncattle

History

The Eastern Finncattle (EFC) was recognised as a separate breed in the 1890's. The EFC farmers founded a breed society in 1898, which initiated organised cattle breeding in Finland. First, attention was given to breed characteristics and the cows from peripheral villages were regarded as the most pure ones. Then there was a need to improve milk production. From the 1920's onwards the emphasis on exterior traits made way for selection on recorded production. The EFC populations sank to the bottom lowest numbers in the 1980's with only about 50 cows and less than 10 bulls left. At the moment the number of purebred cows is almost 800 and slowly increasing. Animals are typically red colour-sided with a broad winding white band on the back.

Breeding, conservation and promotion

The proportion of recorded cows is 32%. The AI organisation has semen stored from 48 bulls with a total of 75,000 doses. There are also 100 embryos in the cryobank produced from 18 cows (12 bulls). The breeding organisation FABA Service annually lists alternative bulls for each cow, recommended on the basis of overall co-ancestry measures in the population. Since joining the EU, the farms raising EFC cows have received a special subsidy.

SWOT

S: Unique and symbolic germ plasm in Finland.
W: Low milk yield.
O: Special features exploited in product development; 'green care' farms.
T: Less experienced farmers and hobby farmers have no interest in the development of milk production.

arise from variability at the molecular level. The fast development of molecular genetics has made a detailed study and quantification of genetic diversity possible (c.f. Woolliams and Toro, 2007). The genome has potentially hundreds of thousands sites showing variation. These could be revealed by different types of markers. Until quite recently the microsatellites have been most popular. But now Single Nucleotide Polymorphisms (SNPs) have become very popular, because SNPs are widely but densely distributed over the genome and cheap and high throughput DNA chips are now available. SNPs occur approximately every 700 base pairs (bp) in *Bos taurus* and every 300 bp in *Bos indicus* cattle (The Bovine HapMap Consortium, 2009), which means there is more genetic variation in *B. indicus* cattle. Thus, there are approximately 4 million SNPs in the *B. taurus* genome. Simultaneously up to 500,000 sites could be typed in one analysis at a cost of €200-300.

Classical measures of genetic diversity calculated from the knowledge of markers' genotypes include some important parameters. The first measure is the *allelic diversity*, defined as the number of different variants that individuals of the population carry in that particular site. This is an intuitive and direct measure related to the long-term potential (capacity) of the population to evolve; the higher the number of variants, the more probable that the individuals can successfully face a new challenge. Thereby the measure is related to quantitative genetic variability of the population. There may be alleles in one breed that are completely missing in the other breeds. With respect to the breed, those kinds of alleles are called private alleles, indicating the uniqueness of the genetic constitution (Groeneveld *et al.*, 2010).

The second measure of genetic diversity is the *observed heterozygosity*. Domestic animals are diploids and every individual carries pairs of alleles at each locus. In homozygotes both alleles are equal, while heterozygotes possess variation by carrying two different alleles. Reduced *observed heterozygosity* is related to the risk of inbreeding. Inbreeding depression is the decline in the performance of a trait (especially fitness) due to the expression of deleterious recessive alleles in homozygotes. The lower the *observed heterozygosity*, the higher the effect of inbreeding depression.

The third relevant measure is the *expected heterozygosity*. Sometimes the *observed heterozygosity* is smaller than the expected one, because the mating has deviated from random. The *expected heterozygosity* is the proportion of heterozygotes that we could see in an ideal population (i.e. no selection, random mating, equal viability of offspring) with the same allelic frequencies as in the actual population. This measure is connected to the ability of the population to respond to selection in the short term. The rate of response depends (among other factors) on the amount of additive genetic variation in the trait. Additive variance is proportional to the

Farmer's quote:

'I don't want to live like Robinson Crusoe; my cows are production animals, not pets'

heterozygosity, being both maximal when all alleles are at the same frequency. This point is relevant to local breeds that are not (yet) at an extreme degree of endangerment, because we may want to perform some selection to make the breed more profitable and, then, more self-sustaining.

As mentioned before, there is an active development of new molecular genetic technology allowing for high-throughput genotyping. The dense marker maps cover the entire genome giving a detailed picture of the genetic variability. This technology has also facilitated the detection of important regions or loci with adaptive effects (e.g. Toro and Mäki-Tanila, 2007). The variation of such genomic regions is especially worth keeping. The loci could be studied further over breeds and individuals using a technique called re-sequencing (e.g. Sellner *et al.*, 2007). As the technology in molecular genetics advances, it is very likely that sequencing of the whole genome of individuals will soon replace the marker typing. This would result in increments in the accuracy of the estimation of genomic variation and, correspondingly, in the power of strategies devoted to the management of the genetic diversity (and also in selection efficiency).

Over generations, the alleles at different loci are recombined. If population size has stayed large over a long period, there has been time to produce recombinations even over a very narrow genome area. On the other hand, in a very small population, the variants tend to be transmitted over longer genome stretches. Such blocks would therefore indicate a small population size (bottleneck) in the recent history of the population (e.g. Toro and Mäki-Tanila, 2007).

When we are studying several populations, the allele frequency differences could also be used to quantify the relationship (through the calculation of different genetic distances) between two populations (Caballero and Toro, 2005; Toro *et al.*, 2009).

6.3.2 Use of pedigree information

Genotypes of a particular marker provide direct information on variability at the locus and frequently at the closely linked loci. When the marker map is not very dense, we may get a biased measure of the diversity of the non-genotyped part of the genome. An alternative way to proceed is to use measures of diversity calculated from genealogical information. Pedigree-derived parameters provide expectations across the whole genome, allowing the estimation (and maintenance) of variability at unknown sites. One of the most important measures that genealogical analysis allows is the probability of Identity By Descent (IBD). It is the probability that two alleles are copies of a single allele in an ancestor.

The probability of an individual carrying IBD alleles is called the inbreeding coefficient (F). Inbreeding leads to homozygosity, and the reduction in the *observed heterozygosity* in the present generation compared to the reference generation is proportional to the inbreeding level F of the current individuals. The probability of a pair of individuals carrying IBD alleles is called the coancestry coefficient (f). Close relatives originate from common ancestors and, consequently, tend to share genetic information. The average coancestry of the population is the complement of the *expected heterozygosity* in the population (i.e. $f = 1$ - *expected heterozygosity*). When relatives are mated, their progeny have high levels of inbreeding (we say they are inbred) as the inbreeding coefficient of an individual is equal to the coancestry of its parents.

The level of inbreeding greatly depends on the depth of the pedigree information that is available. One would expect that the inbreeding coefficients and predictions for homozygosity would be higher for a population where the pedigree recording has started earlier, just because we can trace back further in the genealogy to find a common ancestor. Therefore, we have to be careful in comparing populations with differing amount of pedigree information. We can circumvent this problem by thinking in another way. Because inbreeding is inevitable with a finite number of parents and family lineages, we are more interested in the rate of reduction in heterozygosity in the population or, correspondingly, in the *rate of inbreeding* (ΔF). *Rate of inbreeding* is independent of where we have established the reference population or when we have started pedigree recording and is better at describing the dynamics of variation.

In small populations, the main cause of the loss of genetic variation is genetic drift. Genetic drift is the fluctuation of allele frequencies due to finite sampling of parents and gametes within parents. The increased variation in allele frequencies may lead to the fixation of some allele variants and, consequently, to the loss of alleles. Another consequence of drift is the increase in homozygosity with its undesirable consequences.

In ideal conditions the only parameter influencing the effect of genetic drift is the population size. But in real situations some other factors affect the evolution of allele frequencies and inbreeding. To account for this, the *effective population size* (N_e) is defined as the census size of an ideal population yielding the same ΔF or the same increase in variance of frequencies. We define $N_e = \frac{1}{2}\Delta F$, so ΔF arises as a powerful measure of the 'quality' of the management of the population and a useful parameter on which to base future decisions. Breeds could, therefore, be unambiguously compared for the state of genetic variation using ΔF or N_e irrespective of how long we have been recording pedigree. Among the factors affecting N_e the most important are: (1) the numbers of males and females and the mating ratio (e.g. a population of two males and 1000 females is equivalent to one with four males and four females); (2) fluctuation in census size over generations (a population bottlenecking has a strong effect on N_e across generation); (3) unbalanced number of offspring generated from different parents (N_e is inversely related to the variance of the contributions or the family sizes).

In summary, the pedigree analysis gives an expected value of the increase in the homozygosity and decrease in the heterozygosity for the whole genome, while genomic analysis reveals the realisation of such expectations, and could be used to find deviations from the expectations.

The parameters obtained from genealogical analysis provide useful tools for predicting the consequences for a given management scheme or for designing the resources for a conservation programme, where general variability is to be maintained. The use of molecular information (combined with pedigree data or alone) may be most useful for dealing with adaptive variation and to unveil the old history of populations (i.e. before pedigree recording started) (Toro and Mäki-Tanila, 2007). For further reading on genealogical analysis, see Woolliams and Toro (2007).

6.4 Acceptable levels for *Ne* (*ΔF*)

There are a number of approaches described in the literature to assess the 'acceptable rate of inbreeding' or conversely the 'minimum effective population size' to maintain a relatively 'safe' population. Regarding the short-term prevention of inbreeding depression problems, there seems to be a consensus among animal breeding researchers that *ΔF* of 0.5 to 1% is acceptable. Therefore, an effective size of 50-100 could be sufficient to keep a population in a healthy state. Meuwissen and Woolliams (1994) also considered balancing the depression due to inbreeding, which decreases fitness, against the genetic variation available for natural selection, which improves fitness. Depending on the fitness parameters assumed, the critical N_e varied between 50-100 individuals.

When taking into account other criteria (i.e. long-term potential to evolve and accumulation of mutations), the figures should be higher, with the value depending on the assumptions about the mutational model (i.e. the mutational rate and the mean effect of spontaneous mutations). In conclusion, although there does not seem to be a 'school-book' derivation or experiment showing that N_e needs to be 50-100, most approaches point in the same direction, namely that *rates of inbreeding* of 0.5 to 1% are acceptable for livestock populations. Some organisations (e.g. FAO, 1998) often use the *effective population size* to define the level of endangerment.

Breeding programmes for mainstream breeds are focused on achieving significant gains in the trait of interest but the programmes should also deal with the problems associated with the loss of diversity. One way to cope with the situation is an efficient monitoring process to detect undesirable changes in fitness traits that are sensitive to inbreeding depression. However, a more reasonable strategy is to incorporate restrictions on the level of expected coancestry (or inbreeding) in the offspring with the objective of maximising gain (details will be given later in this chapter).

Farmer's quote:

'Tourists cuddle the cows and take photos of the animals; the neighbours don't mind our cows running into their backyard'

6.5 Estimation of *Ne* (*ΔF*)

The concept of *effective population size* usually has an asymptotic meaning in a regular system, and it is more frequently used for predictive purposes rather than for analysing realised genealogies. However, we can still use the concept to understand the relationship with other pedigree tools. In the context of genealogical analysis we can consider the increase in average inbreeding (coancestry) from the founder generation up to a given generation, or the increments can be calculated generation by generation. For a population with regular dynamics over generations and no other factors than genetic drift modifying the variation, the *rate of inbreeding* would fluctuate around a constant value after a few generations (e.g. Woolliams, 2007).

When pedigree information is available, the usual way is to group the animals by cohorts and calculate the average inbreeding and the average pair-wise coancestry for each cohort. A natural way of establishing cohorts is by the year of birth or by sex and year of birth. If the increase has been more or less linear we can calculate the rate of increase in inbreeding or coancestry just by regressing the average inbreeding (coancestry) on years.

6.5.1 Example on Finncattle breeds

The above methodology is demonstrated with data from the Eastern Finncattle. The breed was established over a hundred years ago when there was general interest in breed formation. The Western and Northern Finncattle were established at the same time. The Finncattle breeds were merged in the mid-1940's. However, the interest in the separation of breeds and their maintenance received attention in the 1970's and the three Finncattle breeds were re-established in the 1980's. Over the period of the last 50 years or so the Finncattle breeds have been replaced by the high-yielding Ayrshire and Holstein cattle (Figure 2.2). At the moment there are some 800 Eastern Finncattle cows. The pedigree file was received from the Finnish cattle breeding organisation (FABA Service) and the pedigree contained over 9900 individuals with varying amounts of information on the parents and ancestors. The generation interval is about seven years, the regression gives an annual *rate of inbreeding* of 0.18%, and the *effective population size* is 39 (Figure 6.1). The trend in the relationship coefficients is less smooth because the pedigree information on animals is very variable, as many new herds with Eastern Finncattle animals with no known link to the rest of the population have been registered over the last two decades.

6.5.2 Estimation of N_e with no genealogical information: demographic data

When pedigree records are not available, the basic idea is to use the number of breeding individuals (sires and dams) in the population and the historic records about population size

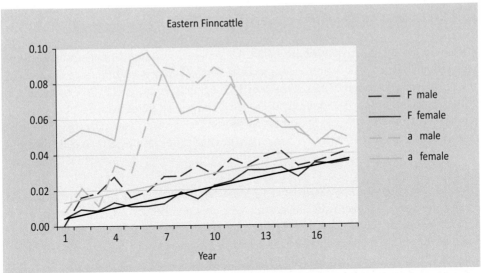

Figure 6.1. Trends in additive relationship (a) and inbreeding (F) in Eastern Finncattle males (dotted lines) and females (solid lines) over a 20 years period. The irregularity in relationships is due to variable pedigree information, as many new herds have entered the animal register recently. The straight lines show the regression of F on birth year in males (grey, y = 0.0018 x + 0.0112) and in females (black, y = 0.0019 x + 0.0024).

and structure. Then the classic formulae for predicting the *effective population size* can be used to account for the deviations from ideal conditions (i.e. unbalanced mating ratio, fluctuating population size, unequal contributions) (Falconer and Mackay, 1996). However, the formula $N_e = 4 S D / (S + D)$, for example, where S and D are respectively the number of breeding males (sires) and females (dams), does not account for selection and leads to an underestimation of expected annual inbreeding rates. In the case of selection (Woolliams and Bijma, 2000) one should avoid the classic formula or aim for N_e, which is at least a factor 3 larger.

6.5.3 Molecular data

If genotypes for a number of markers are available at different times, we can recall the consequences of genetic drift to estimate N_e. One of the effects of genetic drift is the reduction in heterozygosity, and estimates on N_e can be obtained by comparing the number of heterozygotes in two generations. The other consequence of drift is the increase in the variance of the allele frequencies. Molecular data at different times (weighted by the sampling sizes) are commonly used to estimate N_e by means of the so-called 'temporal method' (Waples, 1989).

6.5.4 Example using Spanish beef breeds

As an example Table 6.1 shows both molecular genetic (based on 16 microsatellites) and pedigree analysis of nine Spanish cattle breeds. The *effective population size* is low in all cases (especially for the mountain-type breeds) ranging from 21 to 123. Only three breeds (Asturiana Valles, Bruna del Pirineus and Pirenaica) showed an increase in the inbreeding per generation below 1%, whereas Sayaguesa surpassed 2%. It is worth noticing that there is no clear relationship between *effective population size* and the measures of molecular diversity.

Table 6.1. The observed heterozygosity (H_o), expected heterozygosity (H_e), average number of alleles based on 16 microsatellites (extracted from Cañón et al., 2001), and the effective population size (N_e computed with regression method, (Gutierrez et al., 2003)) based on pedigree information from nine Spanish beef cattle breeds.

	H_o	H_e	No. alleles	N_e
Alistana	0.629	0.681	6.9	36
Asturiana Montana	0.652	0.705	6.6	35
Asturiana Valles	0.656	0.683	7.0	89
Sayaguesa	0.654	0.707	6.4	21
Tudanca	0.596	0.651	6.8	35
Avileña-Negra Ibérica	0.589	0.692	6.9	40
Bruna del Pirineus	0.619	0.672	7.1	95
Morucha	0.640	0.709	6.9	27
Pirenaica	0.543	0.628	5.8	123

6.5.5 Software

There are now software packages available to compute diversity parameters on pedigree data. Many of them produce inbreeding coefficients, with the emphasis on having efficient algorithms for very large pedigrees, e.g. PEDIG (Boichard, 2002) and RelaX2 (Strandén and Vuori, 2006). A more complete treatment, including different methods to estimate N_e, is found in packages like ENDOG (Gutierrez and Goyache, 2005) and POPREP (Groeneveld *et al.*, 2009). Before the pedigree data can be used for the analyses, it should be checked for duplicate IDs, mismatching birth dates etc., and individuals without parent information should usually be added to the top of the list. Some software packages even require renumbering of animal IDs.

Alistana-Sanabresa

History

In the Spanish Breeds Official Census of 1979, Alistana-Sanabresa was considered as two different breeds: Alistana and Sanabresa. Just two years later, Alistana-Sanabresa appeared as a single breed in the Autochthonous Spanish Breeds Catalogue. The breed has a triple aptitude; it was used as a draught animal for agriculture labour, and its meat and milk was also consumed. Nowadays, the breed is dedicated to beef production. In the last decade the breed census seems to be recovering, to >2,500 animals. Alistana-Sanabresa animals can be found nowadays in most of Zamora and in many others provinces of Castilla y León.

Breeding, conservation and promotion

In the 70's selected males were bred to deal with the shortage of Alistana-Sanabresa stub bulls. In 1995, a subsidies programme began to support endangered autochthonous cattle breeds. Three years later the Alistana-Sanabresa Pedigree Book was created. In the same year, a breed association was formed and from that date on they coordinated Alistana-Sanabresa farmers for the development and conservation of the breed. Alistana-Sanabresa farmers usually have their own bulls. In 2007 there were around 500 bulls available. There is no semen conservation programme, and a breeding strategy was non-existent. Recently, a breeding programme for beef production has started.

SWOT

S: High quality of breed products (meat) and low comparative cost of inputs.

W: High dependency on subsidies and lack of data and studies about the breed characteristics and its singularities.

O: Increase in demand for environmental friendly cattle farming and in demand for landscape and vegetation management of extensive cattle farming.

T: Rural depopulation and presence of many other local breeds in the region (North-West Spain).

6.6 Management of genetic variation

The maintenance of variation is related to the *effective population size* or *rate of inbreeding*. From the definition of N_e itself and the factors maximising N_e (or minimising the genetic drift), some basic recommendations can be extracted. First, we should obviously keep the highest possible number of parents and try to have the same number of sires and dams. Then we should try to equalise the number of offspring (contributions) to be obtained from every potential parent. The idea behind this is to give the same opportunity to every parent of effectively transmitting their alleles. And finally, we should prolong the generation interval as genetic drift occurs always when parents' alleles are sampled in creating offspring. Note that this last recommendation and that of using many parents decreases the annual rate of response in a selection scheme.

6.6.1 Hierarchical methods

In practice, in many livestock species it is impossible to reach the 1:1 sex ratio. To cope with this situation some hierarchical (several dams mated to each sire) and regular systems have been developed (Gowe *et al.*, 1959; Wang, 1997; Sánchez-Rodríguez *et al.*, 2003). The idea is always to equalise the contributions from each individual to the next generation (or to all the following generations in the last and most sophisticated method). Basically, these strategies consist of a more or less optimised form of within-family selection.

Hierarchical methods have the advantage of being simple and easy to implement for non specialised personnel and of providing predictions on the evolution of inbreeding over the years. The disadvantage of them is that they are very sensitive to deviations from the assumed conditions (i.e. related founders, mating failures, number of females not being an exact multiple of the number of males, fluctuating population size) as shown by Fernández *et al.* (2003) and, therefore, they are not applicable in most real situations.

6.6.2 Optimum contribution

We know that the coancestry between individuals is directly related to the genetic diversity of the population (measured as the *expected heterozygosity*) and also related to the expected inbreeding in the next generation. Coancestry between individuals also reflects the proportion of common genes and, thus, the redundancy of the alleles in the individuals. From this it follows that a good methodology should consist of finding the combination of contributions from available parents to minimise the expected average coancestry in the next generation. This is achieved by applying the Optimum Contribution strategy (OC) (Meuwissen, 2007). The driving principles would be to penalise the parents closely related to the rest of the population by allowing them to have only a few offspring, while non-related parents will be favoured as they would probably carry a unique variation.

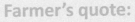

Farmer's quote:

'Be more vocal about the financial results, then all farmers would keep local breeds'

The use of OC methodology is not exclusive to conservation programmes. Long-term selection schemes also benefit from it by including a second term restricting the average coancestry to a desired level in the usual objective function (with a negative sign), directed at maximising the gain (Meuwissen, 1997). There are interesting similarities behind the two terms. In finding the best candidates for selection, the comparison of genetic values also resorts to the information from relatives. The well-known additive relationship matrix used in such an evaluation using the BLUP methodology equals twice the coancestry matrix. In conclusion, with OC selection one can either minimise the *rate of inbreeding* (*ΔF*), or constrain it to a predefined value and maximise genetic gain simultaneously.

6.6.3 Software

Recently, software has been developed for choosing the sires and dams and allocating the contributions for them both in conservation and selection programmes. GENCONT (Meuwissen, 2002) is able to perform OC selection for a given *rate of inbreeding*. EVA (Berg *et al.*, 2006) produces a similar kind of outcome but puts cost weights against the coancestry instead of restricting the *rate of inbreeding*.

6.6.4 Mating strategies

Once the parents and the optimal number of offspring from each of them have been decided, we should determine the mating scheme. It should be noted that the optimisation of contributions is the main task in the management of diversity, leaving little margin for any improvement in the mating design. With a one generation horizon, the genetic level and average coancestry do not depend on the way the parents were mated. Inbreeding is greatly influenced, because the inbreeding of the descendants is, by definition, the coancestry between the mating pairs. If we are worried about inbreeding, it is sensible to implement strategies that prevent matings between close relatives (Meuwissen, 2007). In a general non-regular population, this methodology is called the minimum coancestry mating and consists of finding the combinations of couples that yield the minimum average coancestry between each pair of individuals to be mated. As pointed out by some authors (e.g. Crow and Kimura, 1972), the prevention of mating between relatives is not the best method in the long term, but the method they proposed implies a large increase in inbreeding in the short term, which would not be acceptable in most conservation programmes.

Other strategies like compensatory mating (Caballero *et al.*, 1996) have been proposed. This methodology works by mating the most related females with the least related males, and *vice versa*, trying to balance the genetic contributions from under- and over-represented lineages. However, performances are not really very different from that of the minimum coancestry mating, so the former may be recommended. Henryon *et al.* (2009) proposed mating animals by minimising the covariance between ancestral contributions (MCAC mating), showing that lower levels of inbreeding can be reached when performing truncation selection.

When physiologically feasible, some authors (see, for example, Sorensen *et al.*, 2005) have proved that performing a factorial mating design (i.e. mating each parent to several mates) would reduce the levels of inbreeding achieved due to the reduced correlation between the contributions of mates. Moreover, factorial mating increases the flexibility in breeding schemes for achieving the optimum genetic contributions.

Sometimes, for practical reasons (e.g. a female is not able to mate with more than one male), and results from the OC methodology cannot be fitted into a realistic mating design. In that situation, we would like to determine at the same time not only how many offspring an individual animal should have, but also with which animal it should be mated. The simultaneous optimisation of selection and mating is called 'mate selection' and, instead of deciding just on the number of offspring to be had from each candidate, it also looks into the number of offspring produced from every possible couple. It is easy to include some restrictions on the number of matings per particular individual animal or the maximum number of full-sibs to generate among the progeny. Thorough explanations on how to perform the mate selection can be found in, for example, Fernández and Caballero (2001).

6.6.5 Strategies when pedigree information is missing

When no pedigree is available there are two options. To begin with, we could use molecular information to complete or replace the genealogical information. In its simpler form, it is very common to carry out a paternity analysis, useful for determining the probabilities for the sire candidates (and sometimes also for the dams) in free range animals, and consequently filling the gaps in the pedigree. In more complex situations, we could determine the general relationships in a group of individual animals through a set of available coancestry estimators (for a review see Butler *et al.*, 2004; Fernández and Toro, 2006; Oliehoek *et al.*, 2006). Once

Farmer's quote:

'Because of the good features of the local breeds, farmers save money'

the IBD matrix is constructed, we could use the aforementioned management methods directly. Fernández *et al.* (2005) studied the accuracy of molecular coancestry in maintaining the genetic diversity in a conservation programme when replacing or complementing the genealogy with molecular genetic information. The pedigree based method was found to be very powerful and the number of required markers to mimic the genealogical information was high. The study relied on the use of microsatellites and conclusions should be re-evaluated in the context of dense SNP maps.

The genomic information could also be utilised for comparing the genetic value of individual animals for quantitative traits. The pedigree-based relationships can be augmented or even replaced by marker-based information. This is probably easier to envisage by considering the new genomic selection method (Meuwissen *et al.*, 2001), where the genetic value of an individual animal is determined by summing the effects of tens or hundreds of thousands markers over the whole genome. The marker effects are estimated from a sufficiently large reference population. The genomic selection automatically leads to differentiated selection of relatives (Daetwyler *et al.*, 2007) and opens avenues for maintaining variation in targeted genome areas (Roughsedge *et al.*, 2008). Management of variation is very important in genomic selection because, as a very efficient method, it is expected to lead to long-term depletion of variation with a higher risk than conventional methods (Hayes *et al.*, 2009).

6.6.6 Rotational system

Another way to circumvent the problem of missing genealogies is to implement a regular system of management like circular mating. This is relevant if a non-pedigree local breed is made of geographically isolated herds or sub-populations. In circular mating, the males of sub-population *i-1* mate each generation to the females of sub-population *i*, the males from sub-population *i* to the females of sub-population *i+1* and so on (e.g. Colleau and Avon, 2008). This methodology is simple to implement but has the inconvenience of forcing inbreeding in the short term, as mating is restricted to neighbours and, consequently, some degree of sublining is produced. Obviously, in the long term this strategy is good, because the variants within sub-populations are not at risk of being lost via segregation and in the whole population high levels of *expected heterozygosity* are maintained.

6.6.7 Cryopreservation

Cryopreservation is a very useful tool in the management of genetic variability, and has several advantages: (1) it enlarges the population size, as we can use post-reproductive individual animals and we (can) have a large number of potential parents; (2) it prolongs the generation interval, as individual animals are available for longer periods (recalling the undesirable effect on annual genetic gain); and (3) it acts as a 'back-up' for the genetic constitution of the population, so a population can be recovered after some catastrophic events (Meuwissen, 2007). If only males can be cryopreserved, it is recommended to preserve semen from two generations of males instead of just one founder generation, since this also preserves genes from female founders (Sonesson *et al.*, 2002).

Farmer's quote:

'Farmers with real passion for the breed choose to support biodiversity and opt for high-quality products and low environmental impact'

Criteria for choosing the donors to the germplasm bank and the number of samples to get from each of them should be also based on the minimisation of coancestry (Fernández *et al.*, 2006), as the final aim of the germplasm bank is to store the highest levels of genetic variability. This methodology also enables the interest in maintaining/removing particular genetic information, accounting for the expression of a particular trait, to be combined with the general objective of maintaining genetic variability for the whole genome.

6.7 Conclusions

- Understanding and management of genetic variation should be integrated in the development and utilisation of breed diversity.
- Census and pedigree recording are important.
- The state of genetic variation is best expressed by the *rate of inbreeding (ΔF)* rather than the level of inbreeding.
- When molecular and genomic information is compared with genealogical information, it turns out that:
    - small number molecular markers are not very useful (paternity);
    - genomic information may soon replace pedigree information;
    - genomic and pedigree analyses have different time perspectives.
- Management of genetic variation aims to maximise *effective population size* (minimise rate of inbreeding).
- In the management of variation, choosing parents is more important than mating strategies.
- Optimal methods, such as OC selection, are the best tools for management of genetic variation, but require expertise - simple hierarchical methods could be used first.
- Cryopreservation is a cost-efficient back-up for genetic variation.
- Software exists for analysis and management of genetic diversity.

References

Berg, P., Nielsen, J., and Sørensen, M.K., 2006. Computing realized and predicting optimal genetic contributions by EVA. Proceedings of 8th World Congress on Genetics Applied to Livestock Production, Belo Horizonte, Brazil.

Boettcher, P. and GLOBALDIV Consortium, 2010. Objectives, criteria and methods for priority setting in conservation of animal genetic resources. Animal Genetics (in press).

Boichard, D., 2002. PEDIG: a FORTRAN package for pedigree analysis suited for large populations, in: Proceedings of 7th World Congress on Genetics Applied to Livestock Production, Montpellier,19-23 August 2002, INRA, Castanet-Tolosan, France, CD-Rom, comm. No. 28-13.

Butler, K, Field, C, Herbinger, C.M. and Smith, B.R., 2004. Accuracy, efficiency and robustness of four algorithms allowing full-sibship reconstruction from DNA marker data. Molecular Ecology 13: 1589-1600.

Caballero, A. and Toro, M.A., 2005. Characterization and conservation of genetic diversity in subdivided populations. Phil. Trans. R. Soc. B 360: 1367-1378.

Caballero, A., Santiago, E. and Toro, M.A., 1996. System of mating to reduce inbreeding in selected populations. Animal Science 62: 431-442.

Cañón J., Alexandrino, P., Bessa, I., Carleos, C., Carretero, Y., Dunner, S., Ferran, N., Garcia, D., Jordana, J., Laloë, D., Pereira, A., Sanchez, A. and Moazami-Goudarzi, K., 2001 Genetic diversity measures of local European beef cattle breeds for conservation purposes. Genetics Selection Evolution 33: 311-32.

Colleau, J.J. and Avon, L., 2008. Sustainable long-term conservation of rare cattle breeds using rotational AI sires. Genetics Selection Evolution 40: 415-432.

Crow J.F. and Kimura, M., 1972 An introduction to population genetics theory. Harper and Row; New York.

Daetwyler, H.D., Villanueva, B., Bijma, P. and Woolliams, J.A., 2007. Inbreeding in Genome-Wide Selection. Journal of Animal Breeding and Genetics 124: 369-376.

FAO, 1998. Secondary Guidelines on the Management of Small Populations at Risk. UNEP, Rome.

Falconer, D.S. and Mackay, T.F.C., 1996. An Introduction to Quantitative Genetics, 4th edition. Longman, Harlow.

Fernández, J. and Caballero, A., 2001. A comparison of management strategies for conservation with regard to population fitness. Conservation Genetics 2: 121-131.

Fernández, J., Toro, M.A. and Caballero, A., 2003. Fixed contributions designs versus minimization of global coancestry to control inbreeding in small populations. Genetics 165: 885-894.

Fernández, J., Villanueva, B., Pong-Wong, R. and Toro, M.A., 2005. Efficiency of the use of molecular markers in conservation programmes. Genetics 170: 1313-1321.

Fernández, J and Toro, M.A., 2006. A new method to estimate relatedness from molecular markers. Molecular Ecology 15: 1657-1667.

Fernández, J., Roughsedge, T., Woolliams, J.A. and Villanueva, B., 2006. Optimisation of the sampling strategy for establishing a gene bank: storingPrP alleles following a scrapie eradication plan as a case study. Animal Science 82: 813-821.

Fisher, R.A., 1958. The genetical theory of natural selection. Dover Publ. 13th Ed. Hafner, NY.

Gowe, R.S., Robertson, A. and Latter, B.D.H., 1959. Environment and poultry breeding problems. 5. The design of poultry strains. Poultry Science 38: 462-471.

Groeneveld, E., Van der Westhuizen, B., Maiwashe, A., Voordewind, F. and Ferraz, J.B.S., 2009 POPREP: a generic report for population management. Genetics and Molecular Research 8: 1158-1178.

Groeneveld, L.F. and the GLOBALDIV Consortium 2010. Genetic diversity in farm animals – a review. Animal Genetics (in press)

Gutiérrez, J.P., Altarriba, J., Díaz, C., Quintanilla, R., Cañón, J. and Piedrafita, J., 2003 Pedigree analysis of eight Spanish beef cattle breeds. Genetics Selection Evolution 35: 43-63.

Gutiérrez, J.P. and Goyache, F., 2005 A note on ENDOG: a computer program for analysing pedigree information. Journal of Animal Breeding and Genetics 122: 172-176

Hayes, B.J., Bowman, P.J., Chamberlain, A.J. and Goddard, M.E. 2009 Invited review: Genomic selection in dairy cattle: progress and challenges. Journal of Dairy Science 92: 433-43.

Henryon, M., Sørensen, A.C. and Berg, P., 2009 Mating animals by minimising the covariance between ancestral contributions generates more genetic gain without increasing rate of inbreeding in breeding schemes with optimum-contribution selection. Animal 3: 1339-1346.

Meuwissen, T.H.E., 1997. Maximizing the response of selection with a predefined rate of inbreeding. Journal of Animal Science 75: 934-940.

Meuwissen, T.H.E., 2002. GENCONT: An operational tool for controlling inbreeding in selection and conservation schemes. Proceedings of 7th World Congress on Genetics Applied to Livestock Production, Montpellier,19-23 August 2002, 33: 769-770.

Meuwissen, T.H.E., 2007 Operation of conservation schemes. In: Oldenbroek, K (ed.) Utilisation and conservation of farm animal genetic resources. Wageningen Academic Publishers, pp 167-193.

Meuwissen T.H.E. and Woolliams, J.A., 1994. Effective sizes of livestock populations to prevent a decline in fitness. Theoretical and Applied Genetics 89: 1019-1026.

Meuwissen, T.H.E., Hayes, B.J. and Goddard, M.E., 2001. Prediction of total genetic value using genome-wide dense marker maps. Genetics 157: 1819-1829.

Oliehoek, P.A., Windig, J.J., Van Arendonk, J.A. and Bijma, P., 2006 Estimating relatedness between individuals in general populations with a focus on their use in conservation programs. Genetics 173: 483-96.

Roughsedge, T., Pong-Wong, R., Woolliams, J.A. and Villanueva, B., 2008 Restricting coancestry and inbreeding at a specific position on the genome by using optimized selection. Genetics Research 90: 199-208.

Sánchez-Rodríguez, L., Bijma, P. and Woolliams, J.A., 2003. Reducing inbreeding rates by managing genetic contributions across generations. Genetics 164: 1589-1595.

Sellner, E.M., Kim, J.W., McClure, M.C., Taylor, K.H., Schnabel, R.D. and Taylor, J.F., 2007 Board-invited review: Applications of genomic information in livestock. Journal of Animal Science 85: 3148-3158.

Sørensen, A.C., Berg, P.and Woolliams, J.A., 2005 The advantage of factorial mating under selection is uncovered by deterministically predicted rates of inbreeding. Genetics Selection Evolution 37: 57-81.

Sonesson, A.K., Goddard, M.E. and Meuwissen, T.H.E., 2002 The use of frozen semen to minimize inbreeding in small populations. Genetical Research 80: 27-30.

Strandén, I. and Vuori, K., 2006. RelaX2: pedigree analysis program. In: Proceedings of the 8th World Congress on Genetics Applied to Livestock Production, August 13-18, 2006, Belo Horizonte, MG, Brazil.

The Bovine HapMap Consortium 2009 Genome-Wide Survey of SNP Variation Uncovers the Genetic Structure of Cattle Breeds Science 324: 528-532.

Toro, M.A. and Mäki-Tanila, A., 2007 Genomics reveals domestication history and facilitates breed development. In: Oldenbroek, K. (ed.) Utilisation and conservation of farm animal genetic resources. Wageningen Academic Publishers, pp. 75-102.

Toro, M.A., Fernández, J. and Caballero, A., 2009 Molecular characterization of breeds and its use in conservation. Livestock Science 120: 174-195.

Wang, J., 1997. More efficient breeding systems for controlling inbreeding and effective size in animal populations. Heredity 79: 591-599.

Waples, R.S., 1989 A generalised approach for estimating effective population size from temporal changes in allele frequency. Genetics 121: 379-391.

Wickham, B.W. and Banos, G. 1998. Impact of international evaluations on dairy cattle breeding programmes. Proceedings of the 6th World Congress on Genetics Applied to Livestock Production, Armidale, Australia 23: 315-322.

Woolliams, J.A., 2007 Genetic contributions and inbreeding. In: Oldenbroek, K. (ed.) Utilisation and conservation of farm animal genetic resources. Wageningen Academic Publishers, pp. 147-165.

Woolliams, J.A. and Bijma, P., 2000. Predicting rates of inbreeding in populations undergoing selection. Genetics 154: 1851-1864.

Woolliams, J.A. and Toro, M.A., 2007 What is genetic diversity? In: Oldenbroek, K. (ed.) Utilisation and conservation of farm animal genetic resources. Wageningen Academic Publishers, pp. 55-74.

Chapter 7

Decision-making tools for the development of breed strategies

Daniel Martín-Collado, Gustavo Gandini, Yvette de Haas and Clara Díaz

In this chapter:

- Introduction to decision-making tools useful for developing strategies for local breeds.
- Use of SWOT analysis as a proper instrument to tackle the complexity of conservation of local cattle breeds.
- Contribution of different stakeholders to the development of conservation strategies.
- SWOT analysis of local cattle breeds across countries.
- Application of the acquired knowledge to the development of individual local breed strategies and policies.

7.1 Introduction

The conservation of local cattle breeds is a complex problem. It integrates economic, social, environmental and technical issues, and involves many different stakeholders. Those stakeholders range from direct users, such as farmers and consumers, to indirect users that benefit from local breeds, such as the inhabitants of the farming areas. Some stakeholders may be involved in several economic activities, some of which may compete with local cattle breeding.

The complexity of local cattle conservation also arises from the fact that the environment of local cattle production is dynamic. Agro-ecosystems, functions of cattle, products demanded by society, technologies, etc., are changing all the time. Chapter 2 illustrates the changes over the last few decades relevant to the state of local cattle breeds in Europe.

To solve complex problems we first need to identify and analyse the key driving factors of the system, and to understand how they act and how they can be controlled. The identification

Box 7.1 Strategic opportunities for the Dual-Purpose Belgian Blue breed

Within the EURECA project a SWOT analysis and multi-stakeholders approach was used to define and implement a development process for the Dual-Purpose Belgian Blue breed (DP-BBB). As a starting point, the major factors affecting the current state of the breed were determined. Its main strengths

are the stability of the population size, the well-organised registration in the herdbook and the good features of the animals which are longevity, robustness, calving ease and dual-purpose breed. Major weaknesses are the low number of approved bulls, which could lead to inbreeding problems in the short term, and the lack of differentiated product related to DP-BBB.

The most important stakeholders of the breed (the breeders, the Walloon Breeding Association, the Belgian Blue Breed Herd-Book, the Universities in Belgium and Federal and Regional Authorities) share a major interest in the conservation of the breed, which enabled them to start a multi-stakeholders' collaboration process to develop a conservation strategy. Potential strategies were discussed among stakeholders based on the EURECA analysis of strengths and weaknesses of the DP-BBB and they agreed on the necessity to increase the number of approved bulls and manage the inbreeding. Furthermore, DP-BBB is closely related to the French breed Bleue du Nord which is located on the other side of the border between France and Belgium. Originating from common ancestors, they diverged slightly under differentiated selection objectives, but there is still an exchange of bulls, cows and semen. In order to optimally use and conserve both related breeds in Belgium and France, it was decided to set up a collaboration between breeders and stakeholders of both breeds and to process could also be seen as an exercise of simplification; the problem is dissected into its basic entities in order to understand it and to make decisions. We also need to be aware of the risk of over-simplification; all important factors have to be taken into account in order to avoid undesirable outcomes.

Decisions are based on different criteria whose importance is, consciously or unconsciously, weighted. This raises another key question about the decision-making process: how to ensure objectivity in weighting the different criteria?

This chapter explores the use of decision-making tools for the identification and selection of strategies and policies for the development and conservation of local cattle in the context of Europe departing from an analysis of single breeds.

This chapter also presents the multi-stakeholder approach in two specific field cases: the Dual-Purpose Belgian Blue in Belgium, and the Groningen White Headed cattle in the Netherlands. These field cases are presented in Boxes 7.1 and 7.2, respectively.

---

create a cross-border working group to develop common guidelines for the selection of bull-dams and elite-matings.

It was decided to select 20 bulls each year and to recommend elite-matings in order to either improve production traits or conformation traits, or to regenerate poorly represented old lines of bulls. The creation of the working group allowed a common pool of bulls for both countries to be generated. Furthermore, a common pedigree of both breeds was established, allowing for a study of the relatedness between animals which will improve the inbreeding management. Each year the calves that are born will either enter Artificial Insemination Centres or be sold as natural service bulls. The first male calves were born in late 2009-early 2010. A gene pool is currently being set up, which will in future be supplemented with semen doses from the new bulls.

A joint genetic evaluation of French and Belgian Blue bulls for production traits is also being developed to improve the use of available milk-recording data in both countries. The collection of conformation data was harmonised in both countries and a new scoring grid was established by agreement between all breeders and stakeholders from Belgium and France. This new system is better adapted to the specificities of these breeds. Since August 2009, the conformation of cows has been appraised in the same way. Scores are incorporated in a common database and there are plans to use them in the future in a joint genetic evaluation system.

Breeders participating in this working group and those who agree to have a bull dam are very motivated by this initiative. Some breeders even use embryo transfer in order to maximise the chances of new bulls from the elite-matings. The EURECA multi-stakeholders process has led the Belgian and French breeders and stakeholders of the dual-purpose Blue breeds to a common understanding of the problems and opportunities of their breeds. A large consensus has been established and all breeders and stakeholders are carrying out joint actions aimed at the safeguarding of their breeds.

Box 7.2 Defining development strategies for Groningen White Headed cattle

The strategy development process for the Groningen White Headed cattle started with the definition of the strengths and weaknesses of the breed. The main factors were determined by the Dutch EURECA experts based on farmer interviews and stakeholder consultations, in particular with the

people involved in the Groningen White Headed breed interest group (Blaarkopstichting). The strengths of the breed are related to the following features of the animals: good fertility and strong feet and legs. The weaknesses are partly related to the features of the animals (low milk production), but also to the small population size, and the breeding structure (lack of coordinated breeding programme and breeding goal). The potential strategies to overcome the weaknesses were discussed and two main strategic opportunities were highlighted: (1) to (re-)define the breeding goal and to strengthen and better coordinate the breeding programme and breeding structure, and (2) to develop niche markets for breed-related products, in particular those related to certain production systems (e.g. organic farming or regional specificities) or farm management styles. After further discussion with the main stakeholders (Blaarkopstichting) it was agreed that – in the context of the EURECA project – the primary focus should be on improving the breeding structure and breeding programme. The first steps were to explore individual farmers' production goals and to study the genetic structure of the current population.

7.2 The SWOT analysis: a decision-making tool

Decision-making tools, as part of strategic planning, originated from the business world, more specifically in the American Business School in the 60's (Hill and Westbrook, 1997). These tools were developed to help in the process of making choices in complex systems. One of the most widely used tools is the SWOT (Strengths, Weaknesses, Opportunities and Threats) analysis. In this analysis, factors affecting a particular situation or problem in a company (i.e. breed in our case) are split into internal and external factors. Internal factors refer to the attributes of the company (breed) that can be exploited (strengths) or should be minimised (weaknesses) to achieve a goal. External factors are features that foster (opportunities) or hamper (threats) the performance of the organisation (breed). The two groups of factors also differ by the degree of control that we have on them. External factors cannot be controlled or modified, while internal factors can be managed to alter the current situation.

Strategic decisions can be made based on the analysis of the current and expected future situations by using the SWOT matrix (Weihrich, 1982). The matrix settings help to identify

A questionnaire was sent to 291 farmers, with a response from 111. Most of them were dairy herds with mixed breeds (69%). A minority of the farmers focus mainly on beef production (only 4%), therefore a separate breeding programme for beef production does not seem relevant. The most important traits for the farmers in order of importance were: (1) fertility, (2) calving ease, (3) durability, (4) protein content and (5) somatic cell count. Those are the traits that should be incorporated in a selection index, according to the farmers.

A pedigree analysis was carried out. It showed that the number of calves born per year has decreased sharply since 1980, resulting nowadays in approximately 1000 purebred calves per year. The inbreeding and co-ancestry has increased steadily since 1970; the current rate of inbreeding is 0.48% per generation, which is almost the top limit for a healthy population according to FAO guidelines. An increasing awareness about the limited genetic variation in the current Groningen White Headed population resulted in a collaboration plan between the Gene bank and the Groningen White Headed breeders. A number of breeders are using 'old' semen from the gene bank in order to increase genetic diversity on their farm and in the population.

Based on the results of the farmer survey and genetic analysis of the population, the breed organisation (Blaarkopstichting) decided to continue selection of a sufficient number of bulls putting more value on the five selected traits and minimising the genetic relationships with the current White Headed population. They also initiated a programme to attract 'donor funding' for a revolving fund, in order to be able to increase the number of pure bred AI-bulls available for farmers.

The EURECA SWOT analysis and multi-stakeholders process has led the Dutch breeders and stakeholders of Groningen White Headed cattle to a common understanding of the problems and opportunities of their breeds. The process resulted in a number of concrete actions to strengthen the position of the breed.

interactions between internal and external factors. Strategies can be developed in four ways, as shown in Figure 7.1, (1) to maximise both opportunities and strengths, (2) to minimise weaknesses while maximising opportunities, (3) to maximise strengths while minimising threats, and (4) to minimise both weaknesses and threats.

These four strategies can be defined in more specific terms as follows:
- SO strategy: To use strengths to take advantage of opportunities.
- ST strategy: To use strengths to reduce the likelihood and impact of threats.
- WO strategy: To overcome weaknesses that prevent the pursuit of opportunities, and to make use of the opportunities to overcome weaknesses.
- WT strategy: To be aware of limitations that emerge from the combination of weaknesses and threats.

Avileña-Negra Ibérica – Avilena Negra Iberica

History

In ancient times animals settled in the centre of the Iberian Peninsula and evolved in isolation and dedicated to agriculture labour, which was also important for meat production. In the past, animals were named on the basis of its area of origin. The Spanish 'Avileña' group of herds located in the mountains of Avila Province and neighbour areas resisted to the regressive process that affected the Negra-Ibérica group. In 1980, both groups joined each other in the Avileña-Negra Ibérica breed. The trend of the population is upwards. In 1978, there were 80,000 suckler cows. Eight years later, there were 90,000 suckler cows, and in 2007 the estimated census showed 115,000 suckler cows.

Breeding, conservation and promotion

The Breed Association was created in 1971, and began to deal with the herdbook in 1975. It coordinated the animal performance recording plus the genetic improvement programmes, and it organised markets, meetings, and breed promotion activities. It has also stimulated farmers to develop breed-specific products. In 1990, the Protected Geographical Indicator Label, the 'Carne de Avila' was created and in 2000 a Breed Label emerged.

SWOT

S: Better functional traits than mainstream breed (e.g. robustness, health, fertility, longevity) and a strong historical link between the breed and the territory.

W: Lower productivity and carcass value than mainstream breeds, and therefore lower profitability.

O: Increase the quality of products, and increase the awareness for traditional local product conservation.

T: Increasing input costs.

SWOT Matrix		Internal factors	
		Strengths	**Weaknesses**
External factors	**Opportunities**	**SO Strategy** Maximise both strength and opportunities	**WO Strategy** Minimise weaknesses and maximise opportunities
	Threats	**ST Strategy** Maximise strengths while minimising threats	**WT Strategy** Minimise both weaknesses and threats

Figure 7.1. The SWOT matrix: strategic decisions based on SWOT factors (Weihrich, 1982).

7.3 The use of SWOT analysis for the development of conservation strategies

SWOT analysis has been implemented in the context of the EURECA project. We have adapted it into a specific decision-making tool to assist policy makers, local, regional and national authorities, breeders associations or any other stakeholders, in the identification and selection of strategies for the development and conservation of European local cattle breeds.

The SWOT analysis developed in the context of EURECA has been divided into the following three phases:
1. Definition of the local cattle production system.
2. Identification of driving factors of the system: strengths, weaknesses, opportunities and threats.
3. Identification and prioritisation of strategies for breed development and conservation.

Each phase has specific objectives and depends on the preceding one. In the following sections, the methodology of each phase and the main lessons learned from using SWOT analysis in the EURECA breed cases are presented.

7.4 Definition of the local cattle production system

In the analyses of European local cattle production systems, internal and external factors, and their relationship, were depicted (Figure 7.2). Internal factors include all the aspects that breeders and their associations can control and modify in the short or medium term. External factors are those related to the broader social, political, economic, environmental and technical contexts that breeders cannot control. Following this definition, the European local cattle production system was discussed and defined by partners in the EURECA project.

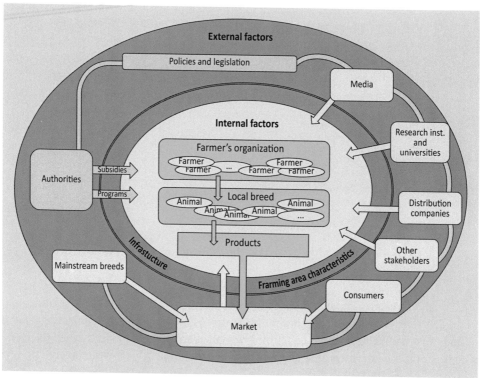

Figure 7.2. European local cattle breed production system as defined in the EURECA project. Internal factors are presented in green and external factors in yellow. Major interactions among factors are represented by arrows.

The production system was simplified to take into account the most important components and their interactions.

Internal factors for a breed (the light green ellipse in Figure 7.2) include farmers, animal, the breed, breed products and their marketing. The farmer factors refer to farmers' characteristics such as age or education, but also aspects related to farmers' organization and collaborations. The animal factors refer to the animals' features, such as level of production or fertility. The breed factors refer to different population characteristics such as demographic trend, effective population size, etc. Products and marketing are also internal factors. Marketing factors comprise those aspects of product marketing that could be controlled by the farmers, like for example, branding or choice of distribution channels. The marketing of products is a key link between internal and external factors, and is represented in Figure 7.2 by the arrow between the Products box and the Market box. The arrows among internal factors refer to all management decisions in relation to animal breeding and transformation of products.

External factors (the blue ellipse in Figure 7.2) include the characteristics of farming area and infrastructure, authorities, policies and legislation, the market and different stakeholders. Among the stakeholders, consumers and mainstream breed farms affect the internal system through the market. Other stakeholders, such as research institutes and universities, media or distribution companies affect local cattle production directly in different ways; e.g. by increasing the knowledge on local cattle breeds, increasing awareness of the society, sales of quality products, etc. Local, regional, national and European authorities probably have the greatest influence on local cattle production of all external factors. They directly affect the local cattle production system through specific policies but they may also regulate its environment. Such regulations affect the activities of the different stakeholders as well as the interaction among them and the market.

7.5 Identification of driving factors: strengths, weaknesses, opportunities and threats

Internal and external factors were identified for each breed considering the opinion of different stakeholders. The breeds analysed were Dual-Purpose Belgian Blue and Dual-Purpose Red and White in Belgium, Eastern and Western Finncattle in Finland, Villard de Lans and Ferrandaise in France, Modenese and Reggiana in Italy, Groningen White Headed, Deep Red and Meuse-Rhine-Yssel in the Netherlands and Avileña-Negra Ibérica and Alistana-Sanabresa in Spain. More information on these breeds is given in Table 4.1. Farmers and other stakeholders were asked about the strengths and weaknesses, and the opportunities and threats for the successful development of their respective local cattle breed.

In total, 108 factors were identified with the following SWOT grouping: 39 strengths, 28 weaknesses, 23 opportunities and 18 threats. These factors were classified for the harmonisation process across countries and across breed cases in order to facilitate a joint analysis of information from different sources and different breeds (see also Chapter 4). Internal factors were classified into the following six categories:
- Animal: productive and functional attributes of animals.
- Breed: aspects related to the population such as size, breed structure, trends, etc.
- Product: quality, uniqueness, etc. of the breed products.
- Farmer: their features, collaboration among farmers, etc.
- Production System: technical, cultural and environmental aspects.
- Marketing System: aspects of the current marketing of breed products under farmers' control, such as branding, distribution channels developed by breeders, etc.

External factors were classified into the following five categories:
- Market for current products: issues related to the market (mainly about consumer demand) for the current products from the breed.

- Market for new products and functions: issues related to the market for new possible products or functions, such us landscape management, tourism, etc.
- Production system: aspects related to the competition with high-input high-output production systems.
- Policies and legislation: regulations ranging from subsidies to health, at regional, national or European level.
- Stakeholders: aspects related to the influence of stakeholders.

The internal and external factors were identified using the opinions of a wide variety of stakeholders and analysing the system from very different perspectives. Table 7.1 contains a list of categories of the stakeholder profiles used.

A consideration of many different stakeholders participating in the process (Table 7.2) would help in identifying a wide range of factors and in detecting all important factors relevant for the analysis. The approach is demonstrated by the identification of weaknesses for the Spanish breed Alistana-Sanabresa. The example is a documentation of a search for different characteristics by a process of considering multiple stakeholders. The outcome of considering different stakeholders would go as follows: farmers and breeders association have identified only weaknesses related to the animal, to the farmer and to the production system. By introducing rural development agencies, the weaknesses concerning products are added to the system. By introducing local cattle farming authorities, the marketing system is considered among the weaknesses of the breed as well. Therefore, value is added to the process by using a wide range of stakeholders in the definition of the key factors, which provides a better understanding of the driving or limiting factors in identifying future strategies for the development of European local cattle breeds. We have introduced more complexity, but also provided a more comprehensive view of the situation. This approach ensures that the system is not oversimplified and that all key aspects are considered.

Table 7.1. Categories of stakeholders used for the identification of SWOT factors.

Stakeholders
National, regional and local Agriculture Authorities
National, regional and local Environmental Authorities
Research Institutes, Universities and State farms
Trade and distribution companies
Rural development agencies
Slaughter houses and dairy cooperatives
Association of breeders and artificial insemination centres
Farmers

Table 7.2. Weaknesses identified by the stakeholders of the Alistana-Sanabresa breed. Green cell connects a factor to the stakeholder who identified it.

Weaknesses groups	Stakeholders					
	Farmers	Assoc. of Breeders	Rural development Agencies	Local cattle farming Authorities	Research Institutes/ Universities	Distribution companies
Animal features						
Product features						
Farmers features						
Production system						
Marketing system						

7.6 Identification and prioritisation of strategies for development and conservation purposes

In the previous section the process of identifying the driving factors of EURECA cattle breeds was presented. The next step is to define and prioritise possible strategies for breed development and conservation. To do so, an expert or a group of experts for each breed ranked the factors within the categories. Experts were used so as to reduce the subjectivity in the ranking of factors. Then, a quantification process was carried out to allow a direct comparison of the different factors of the system. It also allowed us to compare the importance of the different categories of factors and the importance of groups of factors (strengths, weaknesses, opportunities, threats). The quantification process is described in more detail by Martín-Collado *et al.* (2010, personal communication).

The definition of strategies has followed two different approaches that illustrate how flexible the use of the SWOT analysis is. The flexibility makes it a useful tool for a wide variety of stakeholders involved in the conservation of local cattle breeds. First, we made a comparison between the most important SWOT factors identified across breeds, and second, the SWOT matrix has been used to define strategies in different ways.

7.6.1. Comparison of SWOT factors across breeds: lessons from other breeds

The in-depth analysis of the most important internal and external factors (strengths, weaknesses, opportunities, and threats) of each breed can provide useful elements for the management of other breeds. The approach we propose is to study the external conditions (economic, social, environmental and technical) of a given breed that make one of its internal features a strength. If such external conditions are also met for other breeds, one could decide to utilise that specific internal feature to improve their sustainability. As an example, the most important strength of Deep Red cattle in the Netherlands and the Avileña-Negra Ibérica in Spain is the presence of an efficient Breeders' Association. The associations for these breeds in the respective countries have both worked successfully together with farmers and have implemented programmes and actions that led to the success of the breeds. The experience of Deep Red Cattle and Avileña-Negra Ibérica, including the difficulties encountered and the synergies developed, can be shared with other breeds that could benefit their development.

A similar approach can also be used for external factors. What it is about a specific situation that makes it an opportunity for a breed? Could this situation be considered as an opportunity for other breeds as well? Which internal factors have to be taken into account to make use of a recognised opportunity? The threat analysis should focus on becoming aware of factors that are hampering performances in some breeds, and on evaluating the conditions for having the same situation in other cases. Some examples arising from the EURECA breed cases are reported in the following two paragraphs.

Lessons from other breeds' best opportunities

The media can help to increase public awareness of the situation of local breeds. Dissemination of information through the media is the most important opportunity for Eastern Finncattle (Table 7.3). It is useful to analyse the possibilities of involving media to increase awareness of society about local breeds at the regional, national or even European level.

The Groningen White Headed breed has made customers aware of the beneficial features of the animal in particular for low-input and organic farming. This strategy resulted in an increased awareness among the community in the farming area of the positive impact of farming of the local breed on the environment. This aspect can be highlighted to increase the social support to the breed, for example including this aspect in the products' marketing, which might result in increased demand for its products. What other breeds meet similar criteria, and can learn from the experiences of Groningen White Headed?

Traditional products linked to local breeds have proved to be a successful strategy for increasing the price of the breed products and compensating the farmers for the lower breed productivity in comparison to more productive mainstream breeds (Table 7.3). This is the case for the Reggiana breed (Gandini *et al.*, 2007) among others. However, similar cases

La Pie Rouge de Type Mixte – Dual-Purpose Red and White

History

The real selection of Red and White (RW) cattle in Belgium started at the beginning of the 20th century when the colour was fixed by the first breeding associations. The complete standards of the breed were established in 1924. The Dual-Purpose Red and White (DP-RW) had difficulty acquiring as strong a position as the other breeds. Specialised dairy farmers chose Holsteins, and dual-purpose oriented farmers chose Dual-Purpose Belgian Blue. Farmers specialising in meat production preferred the Meat Belgian Blue Breed. Indeed, officially, DP-RW does not exist anymore. Small populations maintained by a few breeders still exist, but they often use foreign bulls, mostly MRY. DP-RW cows are no longer registered in a herdbook dedicated to this breed. The population is declining.

Breeding, conservation and promotion

To date, there are still only a few DP-RW breeders in the Kempen and in the East Cantons but little data is available. Unfortunately, the survival of the DP-RW breed is not assured. There is no herdbook for this breed and no national or regional support for the DP-RW breeders.

SWOT

S: The main strength is the ease of management compared to mainstream breeds.

W: The main weakness is the difficulty in finding different origins of bulls and its negative impact on inbreeding and the lack of support from a breeding organisation.

O: A great opportunity would be the establishment of a herdbook, and the establishment of financial or technical support for DP-RW breeders.

T: The two main threats are the possible end of dairy quotas and the lack of generation transfer illustrated by the high interest among young breeders for the mainstream dairy breeds.

Table 7.3. Most important opportunities for Eastern Finncattle, Groningen White Headed, Modenese and Reggiana breeds.

Breed	Most important opportunity
Eastern Finncattle (Finland)	Increasing media interest in local breed
Groningen White Headed (Netherlands)	Increasing social environmental awareness
Modenese (Italy), Reggiana (Italy)	Increasing demand for products linked to the breed

have to be analysed in detail to learn about the difficulties encountered in the process and the solutions for them; e.g. costs and organisational difficulties associated with the separation of milk collection and milk processing from the mainstream breed farmed in the same area. Unfortunately these experiences are rarely shared, especially where failures are concerned. Can we export the success story of the Reggiana breed to other European local breeds?

Lessons from other breeds' greatest threats

Villard de Lans breed has been negatively affected by other economic activities in the areas where the breed is kept (Table 7.4). This situation is also familiar for many other breeds in Europe, although to a lesser extent than for Villard the Lans. The farmers of European local cattle breeds have to be aware that this kind of threat could be more common in the medium term. Therefore, some defensive strategies could be implemented before the effects become too damaging for breed development.

A second example refers to the increasing demand for Avileña-Negra Ibérica suckler cows for crossbreeding, due to the good maternal features of the cows (Table 7.4). This situation is comparable with other local cattle breeds. Although such demand has a positive effect on the breeders' income, from a genetic point of view it can be a serious threat. Since more and more suckler cows are used for crossbreeding, fewer cows can be kept for pure-breeding, and therefore the effective population size decreases with the associated risk of inbreeding. The use of local breeds with good maternal abilities and fertility as suckler cows to increase breed profitability has to be handled carefully; it requires a sensible breeding strategy to manage genetic variability within the breeds.

Table 7.4. Most important threats to Avileña-Negra Ibérica and Villard de Lans breeds.

Breed	Most important threat
Avileña-Negra Ibérica (Spain)	Increasing demand for suckler cow for crossbreeding
Villard de Lans (France)	Competition against other economic activities

7.6.2 SWOT matrix-based strategies: how to match strength and weaknesses with opportunities and threats?

Once the major internal and external factors have been identified, the next step is to establish useful strategies for the decision-making process. SWOT matrices can be used to identify strategies in four different ways (Figure 7.1). We worked in each breed at two levels: first by matching the groups of internal and external factors, and second, by matching the categories of internal factors with the categories of external factors.

Matching internal and external factors

By jointly considering the relative importance of all the strengths, weaknesses, opportunities and threats within a breed, we can rank the four general strategies (SO, ST, WO & WT) to find the most relevant strategy for the breed (Table 7.5).

Some breeds (Avileña-Negra Ibérica, Ferrandaise, Groningen White Headed and Eastern Finncattle) can focus on strategies that match strengths and weaknesses with the opportunities of their external environment (SO and WO strategies). Others (Deep Red, Modenese, Reggiana and Villard de Lans) can use their strengths to take advantage of opportunities and/or to reduce the impact of the threat (SO and ST strategies). A third group (Dual-Purpose Red and White and Alistana-Sanabresa) has to focus mainly on overcoming weaknesses either to pursue opportunities or to reduce threats (WO and WT strategies). Finally, there are also cases (Dual-Purpose Belgian Blue and Western Finncattle) that are facing such threats that all possible development strategies have to focus on reducing them (ST and WT strategies).

As we have argued above, not all the strategies are of equal importance for all breeds since the current situation among the breeds studied is very diverse. There are differences in the breeds' environment and in the animals' and farmers' basic features such as the level of production, the quality of the products or the farmer's age. Breeds also differ in their 'stages of development'; some breeds are, for example, still in the initial stages of organising a breed association (e.g. Spanish Alistana-Sanabresa), and therefore face the challenges of making farmers collaborate, others have a successful product linked to the breed and the main problem is how to organise the production to address the increasing demand for the product (e.g. Italian Reggiana).

The strategies are mostly breed-specific which should be remembered when discussing the possibility of finding a common general strategy for a group of breeds.

Matching the categories of internal and external factors

More concrete strategies can be highlighted, focusing the analysis on the categories of the SWOT matrix such as animal, farmer, production system, market, legislation, etc. The

Table 7.5. Most relevant general strategies per breed based on the results of the SWOT analysis of the EURECA project. The most highly valued strategy is marked in dark green and the second one in light green (*the four general possible strategies as stated in Figure 7.1).

process consists of developing strategies based on the pair of categories with higher values. In Table 7.6 and 7.7 the quantified SWOT matrix for categories of Avileña-Negra Ibérica and Ferrandaise breeds respectively, are presented as examples. The value of the match between the different categories of internal and external factors - and therefore the value of the subsequent strategy - is shown in a gradient of colours ranging from green (the most highly valued) to red (the least valued).

According to the considered pairs of internal and external groups of factors (Figure 7.1), the general strategy for both Avileña-Negra Ibérica and Ferrandaise breeds should be the same (Table 7.5). However, the analysis of categories shows that when we go one step further in trying to be specific, strategies are more case-dependent. For example, for Avileña-Negra Ibérica, strengths are related to farmer characteristics, but opportunities appear in the market for current products (Table 7.6). On the other hand, the strengths of Ferrandaise refer to breed features, and the opportunities are related not only to the market for new products and functions, but also to the stakeholders (Table 7.7). Therefore, the design of a specific strategy will involve different internal and external aspects depending upon the breed.

Table 7.6. Quantified SWOT matrix for factor categories of Avileña-Negra Ibérica. Colours represent the value of the match of the different categories of internal and external factors, ranging from green (most important) to brown (least important).

Table 7.7. Quantified SWOT matrix for factor categories of Ferrandaise. Colours represent the value of the match of the different subgroups of internal and external factors, ranging from green (most important) to brown (least important).

For Avileña-Negra Ibérica the specific strategy would be to make use of the efficient Breeders' Association to take advantage of the increasing demand for high-quality products and their proximity to important markets (e.g. Madrid). The keepers could try to further optimise the supply chain management from raw material to final delivery, highlighting the quality of the breed's products as a marketing strategy. On the other hand, a potentially effective strategy for the Ferrandaise breed would be to use the increasing interest of the media in the breed to improve the existing genetic conservation plan.

Throughout this chapter we have shown how putting together information about several breeds across Europe in a common SWOT analysis allows us to depict the heterogeneous situation of local cattle production in Europe. We have also shown how SWOT analysis can be used to design general strategies at European level that are based on the identification of breed profiles at a strategic level. We have illustrated how specificity in the decision-making process can be reached by progressing from general factors to more specific categories.

7.7 Conclusions

- The decision-making process for the conservation and development of local cattle breeds requires a clear definition of the system (at breed or breed group level) with all the key driving factors, their internal and external nature and their interactions.
- Definition and prioritisation of strategies need to be based on an objective evaluation of the internal and external factors involved. The multi-stakeholder perspective is an appealing approach to performing a correct weighting of driving factors to help in identifying the best strategies.
- Strategic planning requires working at different levels. We need to go from general to specific strategies and then to decide specific actions and activities for implementing the strategy.
- To increase the effectiveness of common policies at a national or European level, policy makers have to be aware of the heterogeneous state and reality of European local breeds. Different policies and programmes need to recognise different groups of breeds, based, among others, on their 'development stage' and their basic features.
- Common policies have to be extensive enough to be adapted to the specific cases when reaching a lower level, i.e. national level in the case of European policies or regional level in the case of national policies.
- Decision-making tools provide a proper framework for approaching the development of both conservation and breeding programmes in local cattle production and their accompanying policies in Europe. This is possible because they allow a structured and systematic view of the decision-making process.

References

Gandini, G., Maltecca, C., Pizzi, F., Bagnato, A. and Rizzi, R., 2007. Comparing local and commercial breeds on functional traits and profitability: the case of Reggiana Dairy Cattle. Journal of Dairy Science 90: 2004-2011.

Hill, T. and Westbrook, R., 1997. SWOT analysis: It's time for a product recall. Long Range Planning 30: 46-52.

Weihrich, H., 1982. The TPWS Matrix - A tool for situational analysis. Long Range Planning 15: 54-66.

Maas Rijn IJssel – Meuse-Rhine-Yssel

History

Meuse-Rhine-Yssel cattle (abbreviated to MRY) originates from two regions in the Netherlands; (1) along the rivers Meuse and Rhine, and (2) along the river Yssel. Cattle have been registered at CRV since 1874 and in 1905 the cattle were recognised as a breed when the MRY herdbook was started. Up to the '60's and '70's, MRY represented 25% of the total population of Dutch dairy cows (more than 500,000 cows). Since then, the number of cows has decreased rapidly. In 2004, approximately 14,000 purebred MRY cows were registered. In 2008 the number increased slightly to 15,000 cows.

Breeding, conservation and promotion

In 1994 the breed interest groups MRY-East and MRY-South were founded to promote the interests of the MRY cattle. The aim of the breed organisations was to select more MRY bulls to ensure a broader genetic base. Both organisations actively promote the strong characteristics of the breed, and are also involved in developing the breeding programme for MRY with the largest cattle breeding organisation in the Netherlands (CRV). The breeding goal of the dual-purpose MRY breed focuses on milk (35%), functional traits (25%), conformation (25%) and muscularity (15%). Semen from 176 MRY bulls, collected between the 1960s and 2005, is conserved in the National Gene Bank (CGN).

SWOT

S: One of the strengths of the MRY-breed is its profitability, as MRY cows realise a higher net profit per 100 kg milk than Holstein Friesians.

W: The most important weakness is the decrease in genetic diversity within the population.

O: An opportunity for the breed arises from the new interest among the current generation of young farmers looking for more robust cows.

T: A threat for the MRY cattle, on the other hand, is the abolishment of the milk quota in the EU, and the associated trend towards more efficient milk production per cow.

Chapter 8

Recommendations for the management of local cattle breeds in Europe

Sipke Joost Hiemstra, Asko Mäki-Tanila and Gustavo Gandini

In this chapter:

- Discussion of the main results in the context of the objectives of the EURECA project.
- Recommendations for organisations responsible for the development and conservation of local cattle breeds in Europe.
- Recommendations for EU and national policy development.

EURECA objectives

The aim of the EU GENRES co-funded EURECA project was to develop *Guidelines,* in order to support the development of strategies and policies that make local breeds more self-sustaining. A valuable amount of new information, success factors, and factors associated with failures was collected by ten Consortium partners across Europe for sixteen different breeds. We collected data and views directly from farmers, experts, and a variety of stakeholders. From the interaction among actors, we identified new elements and we also tested decision-making processes. We now better understand the similarities and differences between local cattle breeds across countries, factors affecting the demographic dynamics of local cattle breeds, and success factors to make breeds more self-sustaining. In the paragraphs below we discuss the ten most relevant issues and findings, each followed by one or more recommendations for breed managers, policy makers or other stakeholders in this field.

8.1 Policies and strategies for local cattle breeds anticipating livestock sector dynamics

Local cattle breeds in Europe have been developed over centuries. Many breeds were formed and defined at the end of the 19th century. The role and state of local breeds were greatly influenced in the last few decades by many driving forces in the livestock sector. In particular, the modernisation and industrialisation of agriculture after the Second World War resulted in the dominance of a few specialised dairy and beef breeds and many local breeds became endangered. Modern society is consuming more animal products than in the past, and primary production with specialised dairy and beef breeds is part of a standardised and efficient food chain that is very much controlled by international commercial operators. Although the mainstream breeds originate from the local breeds, the major factors driving the livestock sector are often a threat to less competitive and marginalised local breeds. On the other hand, the last two decades show how European/national policies and stakeholder strategies can positively influence the future of local cattle breeds. Since European countries committed themselves to international obligations to conserve and sustainably use animal genetic resources (CBD, FAO Global Plan of Action), National Action Plans are being developed. The results of the EURECA project indicate that National Action Plans should include considerations on how to maintain local cattle breeds and how to make them more self-sustaining. The countries should learn from each other how best to utilise such policies. The FAO Global Plan of Action and the Interlaken Declaration are encouraging countries to pay attention to sound national strategies for local breeds.

Recommendation: Continued development of policies and strategies on how to make breeds more self-sustaining should be included in the National Action Plans for Animal Genetic Resources.

8.2 Monitoring the state of local cattle breeds in Europe

At European and global level, the risk status of local cattle breeds is monitored using information in the EFABIS or DAD-IS databases. A new database network (FABIS-net) has been developed recently, integrating the world database DAD-IS, the European regional database EFABIS and country databases. National Coordinators for Animal Genetic Resources regularly update breed information and census data in these databases. However, the amount of breed data and the quality of specific breed and census data is often poor. Demographic information is often not very precise and there is hardly any information about non-demographic factors affecting the risk status. Several criteria have been proposed internationally to measure the state of endangerment of breeds. In recent years several countries have developed their own criteria and risk thresholds. Via the EURECA project we learned that there are a large number of (non-demographic) factors that can influence breed endangerment status. Most of the factors identified are rather breed-specific and not easy to compare across breeds and countries, and in many cases information collection and trend monitoring is expensive. Nevertheless, there is a need for additional, simple and common indicators to assess the state and trend of local cattle breeds. The results of the EURECA project show, for example, that (1) age of the farmer, (2) degree of entrepreneurial activity of the farmer, and (3) relevance of tradition as the main reason behind keeping the local breed, are relevant factors determining the 'risk status' of breeds in addition to breed census data.

Recommendation: Indicators to assess the risk status of local cattle breeds should contain both demographic and non-demographic information. There is a need for more research in this area and as a result, some (non-demographic) factors should be included in international breed databases.

8.3 Who is keeping local cattle breeds?

Based on a qualitative analysis performed within the EURECA project, we identified three main types of farmers keeping local breeds, according to farming goals and perspectives: (1) production-oriented, (2) product-oriented and (3) hobby-oriented. The farmers are the

Farmer's quote:

'Local breeds have more character and vigour'

ones who decide on whether or not to keep a local breed. The main reason for collecting data in a farmer survey was to find out the profiles and attitudes among the farmers. We will only be able to maintain local breeds in Europe, if a sufficient number and a variety of farmers are willing to keep the breed in an economically sustainable farming system. For local breed development it is important to utilise the interest among different types of farmers. Some types may be less stable than others, but still very relevant in better establishing the survival and successful future of a breed. Therefore, it is very important that we realise that all measures to make breeds more self-sustaining should be supported or driven by farmers. European legislation allows incentive payments to farmers per livestock unit in order to fill the gap in profitability between local and mainstream breeds. Although such measures will certainly help maintain a sufficient number of animals in a local breed, they may also have adverse effects and may not always be the most efficient approach.

> **Recommendation:** Development of measures to support conservation of local cattle breeds should always take into account the (future) farmers' profiles and attitudes.

8.4 Use of decision-making tools for the development of individual breed strategies

In the EURECA project we used and tested the SWOT (Strengths, Weaknesses, Opportunities, Threats) methodology to assess the position and future perspectives of breeds. SWOT appeared to be useful as a decision-making tool and as a means of tackling the complexity of specific local cattle breed situations. There are different ways to use SWOT on an individual breed basis, at the same time it is important to look at breeds from different angles and different stakeholder and expert perspectives. The SWOT approach is extremely useful in itemising the studied system, e.g. dividing the relevant factors detected into internal (controllable by farmers) and external factors (non-controllable). The EURECA project showed that this methodology for identifying and analysing internal and external factors can be used in a multi-stakeholder context to develop and decide upon the most promising strategic opportunities for a breed. Regarding stakeholders, a Europe-wide survey clearly showed that breed societies, national governments and universities/institutes are the main actors in the conservation of local cattle breeds.

> **Recommendation:** Strategic planning at breed level should make use of decision-making tools, in particular the SWOT methodology, and cover all relevant stakeholders in such an analysis.

8.5 Common policies and strategies

There is a large heterogeneity among European local cattle farming systems and local breeds in Europe. Although we identified a number of common factors across the countries and breeds in the EURECA project, we have to be aware of this heterogeneity when developing effective common policies and strategies at the European (or national) level. Common policies to make breeds more self-sustaining might include measures to: (1) support the transfer of farms to the next generation, (2) promote collaboration among farmers, and (3) raise citizens' awareness about the positive roles of farmers of local breeds for society. Policy development should also recognise different groups of breeds which can be distinguished based on their development stage, their basic features, or the context of their local cattle breed production systems. Common policies have to be extensive enough to be adapted to specific breed cases, their needs, and national or regional specificities. On a European level, a number of policy areas might, directly or indirectly, promote or hamper the use of local cattle breeds. National strategies related to local cattle breeds should take full advantage of the EC Regulation opportunities for specific conservation actions. On the other hand, there is also a need to link local cattle breed strategies (or livestock biodiversity strategies in general) more closely to the main European policy areas. For example, livestock biodiversity and rural development objectives can be easily connected, or strict sanitary measures should not unnecessarily hamper the conservation and use of local cattle breeds.

> **Recommendation:** Common policies should not create unbalanced effects across Europe and should be accompanied by local policies tailored to specific country/breed situations ('one size does not fit all').

8.6 Management of genetic variation within local cattle breeds

Local breeds are full of genetic potential. Understanding and management of genetic variation should be integrated in the development and utilisation of breed diversity. In the context of breed conservation, we should be particularly aware of the risk of a small effective population size (genetic drift and inbreeding) and the risk of indiscriminate cross-breeding. Our survey indicated that the monitoring and control of genetic variation only occurs in some of the breeds. In order to assess genetic variation within populations, pedigree recording is a must. Genomic information from high-throughput analyses may in the long run provide equivalent information for estimating the population and quantitative parameters. After an assessment of genetic variation within a population, it is important to decide on an effective management strategy to keep the rate of inbreeding low and to keep the population vital and evolving. Simple methods to minimise coancestry can be used. The number and use of parents for the next generation is more important than mating strategies. In order to help breed managers to make optimal decisions, the EURECA project showed that there is a need for easy and user-friendly tools and software.

> **Recommendation:** Breed managers should give proper attention to the management of genetic variation in order to avoid high rates of inbreeding and to keep the population vital. Researchers and technicians should further develop user-friendly software and strategies.

8.7 Breeding programmes for maintaining/improving the performance

Competitive breeds are more likely to survive than the ones lagging behind. Therefore, it is important to improve or at least maintain the performance in the best traits of the breed. Many local cattle breeds lack efficient and sustainable breeding programmes. The EURECA project showed that well-organised breeding programmes will contribute to the competitiveness of the breed in the long run. In particular in small populations and when only a small number of good sires are available, planned mating and breeding plans are also needed to prevent inbreeding problems. This is best achieved by optimising the contributions of parents with respect to genetic gain and influence on average coancestry. Technical and organisational support to breed societies will be helpful. Where local breeds are transboundary breeds, the local breed can benefit from cross-border cooperation. Although breeding goals can be slightly different between countries, effective population size will increase when starting cross-border cooperation. Breeding programmes should also aim to value those aspects of the breeds and their farming systems that are not yet recognised by the market, such as cultural, social and environmental roles.

> **Recommendation:** Breed managers should strengthen breeding programmes and optimise breeding policies in order to minimise the productivity or profitability gap with mainstream breeds.

8.8 Cryopreserved gene banks

Long-term storage of semen is relatively easy and cheap for cattle cryopreservation in comparison to maintaining large groups of live animals. Semen from a cryoreserve can be very valuable as a support for breeding programmes of local breeds when the number of good quality bulls in the current population is limited and when there are inbreeding problems. Besides semen, it should be further promoted to store embryos as well, in particular for local breeds 'at risk' or at a critical stage. Countries organise cryopreservation programmes at the national level differently, depending on the role and responsibilities of different stakeholders. However, in all four countries (Finland, France, Italy and the Netherlands) studied in detail in the EURECA project, the close involvement of breeders of local breeds, breed associations and AI centres in linking the cryopreservation schemes with routine AI operations is an important factor for the development of efficient cryopreservation programmes. Furthermore, European and national (sanitary) legislation should facilitate and not hamper cryopreservation.

> **Recommendation:** Countries should (further) develop cryopreservation within the framework of a national programme for AnGR. Besides the use of national cryopreserved stocks to support genetic management of local breeds, trans-national cooperation should be encouraged to avoid duplicates in long-term cryoreserve of transboundary breeds and in order to make optimal use of funds.

8.9 Self-sustaining breeds and sustainable local cattle farming

Since 1992 European farmers of local cattle breeds have benefited from EU economic support, in a form of payments to farmers proportional to the number of animals. The EURECA project observed some dependence of farmers on EU economic support, although it is evident that payments in many cases do not cover the gap in profitability existing between keeping a local and a mainstream breed. We know that in some cases economic support to farmers has probably been effective in halting the decline. However, it is a general opinion that economic support cannot last forever, and that we should aim to make local cattle breeds self-sustaining, as has been mentioned often throughout this book. We define a breed and its farming system as sustainable when it is capable of maintaining the vigour and the potential to fulfil the specific conservation aims, ranging from maintaining the genetic variability to the cultural, environmental and socio-economic values. Those – often breed-specific – conservation aims require specific strategies and support measures. Moreover, there is a need for strict eligibility criteria for subsidies, e.g. compulsory pedigree recording. A new EU agri-environmental programme after 2013 should maintain the powerful support for local cattle breeds and augment the programme with new kinds of developmental tools aiming at higher sustainability.

> **Recommendation:** The process towards independence from economic support should be initiated for all breeds, making an appropriate use of all breed strengths and opportunities.

8.10 The final question: Did our research contribute to making breeds self-sustaining? What did we learn from the EURECA project?

It was clearly a challenge investigating the factors affecting sustainability of a breed among other breeds. The development of methodologies and questionnaires that could be used across countries is a complicated issue. Socio-economic, cultural and other differences make it difficult to find a useful common approach. Moreover, it was also difficult to keep the data collection efficient in terms of manpower and costs. Farmer interviews resulted in new data. Due to obvious cost constraints we had to limit the number of breeds investigated and the number of interviewed farmers per breed. On the other hand, we collected a huge amount of information related to local cattle breeds in Europe. Although it is not always easy to draw firm conclusions across breeds and countries, the EURECA Consortium is convinced that the exchange of experience and knowledge across breeds and countries will certainly inspire people to change or strengthen their policies or breed strategies. Moreover – and very importantly – we noticed an increasing interest in society towards local breeds and for many breeds there is a new spirit among farmers, farming communities and local cattle breeders. This is most visible in a substantial number of initiatives to add value to local breeds, for example by branding their products.

> **Recommendation:** Continue exchanging experiences and knowledge on breed strategies and how to make breeds self-sustaining.

Keyword index

Printed in the United States
by Baker & Taylor Publisher Services